The
Transformation Myth

Management on the Cutting Edge series

Paul Michelman, series editor
Published in cooperation with *MIT Sloan Management Review*

The AI Advantage: How to Put the Artificial Intelligence Revolution to Work
Thomas H. Davenport

The Technology Fallacy: How People Are the Real Key to Digital Transformation
Gerald C. Kane, Anh Nguyen Phillips, Jonathan Copulsky, and Garth Andrus

Designed for Digital: How to Architect Your Business for Sustained Success
Jeanne W. Ross, Cynthia Beath, and Martin Mocker

See Sooner, Act Faster: How Vigilant Leaders Thrive in an Era of Digital Turbulence
George S. Day and Paul J. H. Schoemaker

Leading in the Digital World: From Productivity and Process to Creativity and Collaboration
Amit S. Mukherjee

The Ends Game: How Smart Companies Stop Selling Products and Start Delivering Value
Marco Bertini and Oded Koenigsberg

Open Strategy: Mastering Disruption from Outside the C-Suite
Christian Stadler, Julia Hautz, Kurt Matzler, and Stephan Friedrich von den Eichen

The Transformation Myth: Leading Your Organization through Uncertain Times
Gerald C. Kane, Rich Nanda, Anh Nguyen Phillips, and Jonathan R. Copulsky

The
Transformation Myth

Leading Your Organization through Uncertain Times

Gerald C. Kane, Rich Nanda, Anh Nguyen Phillips,
and Jonathan R. Copulsky

The MIT Press
Cambridge, Massachusetts
London, England

The MIT Press would like to thank the anonymous peer reviewers who provided comments on drafts of this book. The generous work of academic experts is essential for establishing the authority and quality of our publications. We acknowledge with gratitude the contributions of these otherwise uncredited readers.

This book was set in Stone Serif and Stone Sans by Westchester Publishing Services. Printed and bound in the United States of America.

Library of Congress Cataloging-in-Publication Data

Names: Kane, Gerald C., author. | Nanda, Rich, author. |
    Phillips, Anh Nguyen, author. | Copulsky, Jonathan R., author.
Title: The transformation myth: leading your organization through
    uncertain times / Gerald C. Kane, Rich Nanda, Anh Nguyen Phillips, and
    Jonathan R. Copulsky.
Description: Cambridge, Massachusetts: The MIT Press, [2021] |
    Series: Management on the cutting edge | Includes bibliographical
    references and index.
Identifiers: LCCN 2020052176 | ISBN 9780262046060 (hardcover)
Subjects: LCSH: Crisis management. | Organizational change. |
    Information technology—Management.
Classification: LCC HD49.K36 2021 | DDC 658.4/056—dc23
LC record available at https://lccn.loc.gov/2020052176

10 9 8 7 6 5 4 3 2 1

Contents

Series Foreword

The world does not lack for management ideas. Thousands of researchers, practitioners, and other experts produce tens of thousands of articles, books, papers, posts, and podcasts each year. But only a scant few promise to truly move the needle on practice, and fewer still dare to reach into the future of what management will become. It is this rare breed of idea—meaningful to practice, grounded in evidence, and *built for the future*—that we seek to present in this series.

Paul Michelman
Editor in Chief
MIT Sloan Management Review

Introduction: COVID as a Case Study in Leadership

On January 8, 2020, Chinese researchers reported that they had identified a new virus that had infected dozens of people across Asia. Over the following months, this novel coronavirus—later named SARS-CoV-2 that leads to an illness known as COVID-19—swept the world and disrupted the global economic, medical, transportation, and political infrastructure. Countries, states, and regions across the world issued a patchwork of stay-at-home orders, face covering requirements, travel restrictions, closures of nonessential businesses, and limitations on in-person gatherings. For many organizations, it triggered decisions to require employees to work remotely or students to attend virtual classes.

In the weeks and months that followed, organizations undertook significant efforts to respond to threats of infection to employees and customers, supply chain interruptions, demand inflections, and extraordinarily high levels of uncertainty. Although as of this writing the grim statistics from the COVID pandemic lag those of the great flu pandemic of 1918, which infected more than a third of the world's population at the time, the numbers are staggering in and of themselves and are symptomatic of a crisis that few anticipated in the months prior to its onset.

Yet, in researching and writing this book, we realized that the COVID pandemic is just one of many major disruptions that organizations have faced over the past thirty years or so, including, but not limited to, the fall of the Soviet Union in 1991 and the end of the Cold War, the Y2K crisis at the end of the last millennium, the dot-com boom and bust shortly after the turn of the century, the September 11, 2001, terror attacks in the United States, the 2004 Indian Ocean Tsunami, the 2008 housing and financial crisis, the 2010 Icelandic Volcano disruption, and the 2016 Brexit vote.

Furthermore, the response to disruption may magnify or reduce its effects. For example, although the stock market crash of 1929 is often seen as the disruption that led to the Great Depression, it was most certainly exacerbated by other decisions and events that occurred in the subsequent years. Significant policy or regulatory shifts, such as the 2002 Sarbanes-Oxley Act or the 2018 European Union General Data Protection Regulation (GDPR), may also be considered a type of disruption that forces companies to adapt quickly.

In fact, in the United States, we witnessed a compounding effect while working on this book when the death of George Floyd on May 25, 2020, in connection with his arrest by Minneapolis police officers triggered worldwide protests and launched a new awareness about race relations within societies and organizations. What marks these kinds of disruptions is that they are sudden and dramatic, as opposed to longer-term shifts like climate change or the ever-increasing digitization of all aspects of business.

The term *disruption* has acquired an important meaning within the business community. Clayton Christensen employed the term disruption in a very specific way to describe the process by which a smaller company or insurgent focuses on the needs of overlooked customer segments to grow and eventually compete with and overtake larger incumbents.[1] We are not employing Christensen's use of the term here. Instead, we are using the term disruption in the formal dictionary sense—a break or interruption in the normal course or continuation of an activity or process. We view disruption as any fundamental change in the business environment in which an organization operates and competes.

As we studied companies' response to COVID, it quickly became apparent that companies were rapidly adapting to a confluence of multiple, interrelated disruptions that interacted in complex ways. For example, it became clear that organizations' response to COVID was inextricably linked to the longer-term digital disruption that had been ongoing (and that we had studied) for years. The shift to remote work enabled many Silicon Valley tech companies to begin hiring a more diverse workforce because they were not limited to people who wanted to (and could afford to) live in the San Francisco Bay Area, a trend that was also driven by the resurgence of the Black Lives Matter movement following a string of highly publicized incidents of police brutality. When we asked Jacqui Canney, chief people officer of WPP, which of these disruptions were most salient for many of the trends she sees, she replied, "It's all of it. I don't think we'd be as far along as we are if it was just COVID."

Thus, when we use the specific term *COVID disruption*, we are referring not only to the effects of the virus but also the various responses (e.g., shift to remote work and policy responses) and secondary impacts (e.g., economic fallout). Furthermore, while the various impacts of COVID is our primary focus, we also include other disruptions of 2020—such as acts of police brutality and the ensuing protests and companies' experiences with digital disruption and transformation—when they deepen our insights. In doing so, we hope our insights will remain applicable to myriad types of disruption that companies are certain to experience in the future.

The changes in the business environment ushered in by disruption can represent both a threat and an opportunity to businesses that experience them, and we intend to focus on both in this book. The core thesis of our book is that *companies that emerge stronger because of disruption are those that use it as an opportunity for innovation*. Our research suggests that business leaders agree strongly with this thesis; roughly 75 percent of CEOs indicate that the COVID crisis has created significant opportunities for their companies.[2]

What Is the Transformation Myth?

Transformation is the process of companies adapting to changes in the business environment ushered in by these disruptions. As the business environment is fundamentally altered as a result of a disruption—whether the result of an inability to go to the office due to social distancing concerns, changing customer demand and shopping patterns, or new competitive capabilities that result from technological innovation—companies must transform to overcome or capitalize on these changes to remain viable.

The word "transformation" itself is a noun that refers to "a thorough or dramatic change in form or appearance," but it has specific applications in different fields.[3] In physics, it refers to the "the induced or spontaneous change of one element into another by a nuclear process." In mathematics, it is "a process by which one figure, expression, or function is converted into another that is equivalent in some important respect but is differently expressed or represented." In biology, it refers to "the genetic alteration of a cell by introduction of extraneous DNA, especially by a plasmid." In linguistics, it is "a process by which an element in the underlying deep structure of a sentence is converted to an element in the surface structure."

Yet, all the definitions of transformation listed above refer to a linear process that has a before, during, and after. This linear pattern holds when adapting to a single disruptive event. Instead, the multiple and interacting disruptions that we describe here mean that disruption is ongoing and not easily predicted. Therefore, the transformations that companies engage in must be ongoing and seen as an active and continuous process. It is about organizations continually transforming and adapting to a disruptive environment in an effort to make themselves stronger and more competitive. Shamim Mohammad, chief information and technology officer (CITO) of CarMax, poignantly illustrated the point by recalling a 2016 interview he conducted with *Forbes*.

> I said in 2016, "I do not know how the world is going to be in three or four years. It is hard to predict. What I am trying to do as CITO is position our company so that we are ready to take those changes and be nimble, agile, and responsive: an organization that can move quickly. That is what I do, because I cannot predict what is going to happen."

Mohammad explained the relevance of this 2016 quote, saying, "In essence, that is what we've been doing as a company over the last few years. We discovered we were able to respond fairly quickly to the disruptions caused by COVID because of all the things we have done to build the company to respond to change in general."

We use a similar rationale for entitling this book *The Transformation Myth*. The "myth" is that transformation is an event with a start and end whereby organizations migrate from one steady state to another as opposed to a continuous process of adapting to a highly volatile, ambiguous, and uncertain environment shaped by multiple, overlapping disruptions. In observing organizations' responses to the COVID pandemic, it quickly becomes apparent that digital transformation is not a "one-and-done" effort to weather the storm until the organization can go back to normal or even the next normal.

Our primary objective in writing this book is to identify and describe the traits that characterize organizations that not only survive disruption but also manage to thrive within it and the characteristics of the leaders who helm these organizations. We use the term "digital resilience" to describe organizations and leaders whose digital transformations can withstand or recover quickly from the difficult conditions created by disruptive events such as the COVID pandemic.

To be clear, we are not claiming that digital technologies are a panacea when it comes to organizational transformation or learning. Quite the contrary. When speaking with executives, we often emphasize the point that just because an organization's challenges are caused by digital disruption, it does not necessarily follow that the solutions will be digital as well. We consciously try to keep the spotlight on the strategic, talent, leadership, organizational, and cultural aspects of digital resilience that are so frequently overlooked in the face of shiny new digital tools.

Nevertheless, completely ignoring the technological aspects of digital resilience is neither possible nor helpful in the current setting. A robust digital infrastructure has proven invaluable for helping companies adapt to disruption, and certain technological investments have paid off better than others. Mastering both the technological and organizational aspects of digital resilience are critical for navigating the disruption of today's world—particularly one as severe as experienced in the wake of COVID.

The digitally resilient organization is one that is continually sensing, testing, and adapting to find its way forward in a turbulent business environment, requiring both technological tools and organizational capabilities to do so. At times, this necessitates "unlearning" practices and processes that were fundamental to the success of prior "versions" of the organization. For example, the six sigma method of error reduction that was wildly popular a generation ago turns out to inhibit agility and an entrepreneurial mindset that are critical for digital resilience.

About the Research

Our prior book, *The Technology Fallacy*,[4] helps organizations start thinking about how to embark on a digital transformation by focusing on people, not technology. The book assumes an environment of low-level disruption, where business leaders could choose when and how to begin their digital transformations. It introduces the concept of "digital maturity," which is a "flexible process by which the organization can continually adapt to a changing technological environment, realigning its people, culture, tasks, and structure in response." Digital maturity involves developing not only a robust technological infrastructure but also new talent models, cultural characteristics, task definitions, and organizational structures appropriate to the fluid environment that characterizes the digital world. We describe the companies closest to the ideal as "maturing," those farthest from the ideal as "early," and those in-between as "developing." We will retain that terminology when we rely on data from our earlier research in the coming pages.

This second book was inspired by the realization that much of the data collected for and insights generated by our first book are remarkably applicable to the challenges that companies experience with respect to COVID. When talking with executives around the world, we realized the need for greater understanding about how to manage digital transformation when there is no choice in the matter—when transformation is forced upon organizations due to sudden and unexpected shocks. We went back to the data we collected for *The Technology Fallacy* and reinterpreted it through this new lens, which generally provides the foundation for the insights in this book. Digital maturity is certainly related to digital resilience but is not fully equivalent to it. Understanding the similarities and differences is the purpose of this book.

We have five years' worth of research, conducted in partnership with *MIT Sloan Management Review* and Deloitte, into how technology changes the way companies operate. We surveyed more than 21,000 people about their experiences with digital disruption and their perceptions of the nature and adequacy of their organization's response. We spoke with over one hundred executives and thought leaders to learn about their stories and experiences. While this research grounding provides a valuable foundation, we go beyond this data in this book.

We conducted additional research to augment this foundation, interviewing approximately fifty additional executives and thought leaders, between March and October 2020 across multiple industries as they sought to lead their companies through disruption, helping us deepen our insights into what companies were dealing with and what characteristics would help them successfully emerge from the crisis. We draw on different quantitative data collected during the COVID crisis, such as Deloitte's biennial global technology leadership study,[5] other Deloitte studies, and Gensler's US Work from Home Survey 2020,[6] to augment findings from our primary data collections.

Yet, we do not want to be bound entirely by the data. Doing so would leave us with a backward-looking approach, no matter how timely, that simply describes how companies have responded to disruption. Instead, we use this data to make proactive and prescriptive recommendations on how managers should respond. We hope to balance our insights with different types of experience to provide the most holistic picture as well as practical guidance and exercises for how to move organizations forward amid disruption.

When relevant, we augment the insights of our primary research by drawing on established literature in the fields of information systems, management, marketing, psychology, and operations to set our findings in the broader context of management science. By triangulating our data using multiple and dissimilar sources—quantitative and qualitative methods, primary and secondary data—we hope to offset the limitation of any single data source to provide a balanced, authoritative, and novel insight into the problem of how companies effectively respond to disruption. We rely on the work of other business school academics, extensively reviewing the insights and guidance published in top business journals like *Harvard Business Review* and *MIT Sloan Management Review*.

Last, we rely on our advisory work with companies on how they are practically navigating disruption to understand and evaluate possible responses to disruption. Using these experiences, we put together a series of exercises that we have used with companies seeking to strengthen their agility, resilience, and response to disruption. We hope that this book will not only help leaders better understand and navigate disruption but also provide practical tools that will help them do so. We provide three unique kinds of tools:

1. Simulations, where we guide the reader on how to think about a problem
2. Frameworks, where we provide a blueprint (e.g., 2×2) for how to structure the problem and make critical choices
3. Checklists/questions, where we help the reader answer a series of questions to get the outcomes they need

Each chapter will make use of one of these different types of tools, depending on which type of tool is most applicable to the content. We expect and hope that these tools will remain helpful for leaders and organizations to prepare for whatever challenges and disruptions their organizations may face, long after the pandemic is over.

Our author team consists of both academics researching and teaching about digital disruption and strategy consultants who have been actively working with companies to help them adapt to the challenges of digital disruption. We don't want the book to be exclusively about COVID, so we explore how lessons learned from earlier crises can be applied to the current environment as well as extrapolate how the unique lessons learned in this crisis can be applied more generally.

Throughout the book, we tend to refer to our collective experience as "we," even if it represents only a portion of the four-person author team. Each member of the team has contributed different expertise and experiences. For example, we rely extensively on Kane's academic research, conducted with various coauthors, to inform our findings. Nanda, an active Deloitte principal and leader of its strategy practice, works closely with clients on these issues. Phillips led the earlier research project on the Deloitte side, is intimately familiar with the data and findings, and leads ongoing related research from Deloitte. Copulsky, a retired Deloitte principal, a widely published author, and a current faculty member at Northwestern University, sponsored the original research and brings a talent for distilling complex ideas into compelling narratives. Our team worked together remarkably well in the development of this book, and we believe the whole is more than the sum of the individual contributions. Despite the individual roles played by team members, we speak with a singular voice to reflect this integrated approach to the project.

By bringing together different experiential perspectives to our data, we hope to offer a clear voice, supported by research, for how companies should understand and respond to disruption. We hope the result is a balanced, authoritative treatment that provides leaders with actionable guidance about how to lead their organization into the future at whatever level their influence allows.

What to Expect

We divide this book into three different parts. Part I provides a framework for how leaders can understand the nature of the disruptions they face as well as provides tools for helping them lead their organizations through various types of disruption.

- In chapter 1, we differentiate between two different types of disruption—acute and chronic. Our previous book focused primarily on the chronic disruption caused by the widespread use and adoption of digital technologies. We contrast that with the acute disruption that companies deal with in response to events like the COVID pandemic to explore how chronic and acute disruption differ from one another.

- In chapter 2, we discuss the stages that most winning companies go through in the process of coping with disruption. We suggest that companies adapt to disruption through three distinct phases. Companies first *respond* to the disruption, then *regroup* following the onset of the disruption, and then use the disruption as an opportunity to innovate through the disruption and *thrive*. Although we primarily consider how these stages apply to responding to acute disruption, we identify commonalities with chronic disruption as well.

- In chapter 3, we explore the nature of disruption as a crucible of leadership. We explore how leaders can employ purpose, values, and mission in positioning companies to navigate crises successfully. When employees understand why, how, and what the organization wants to accomplish, crises can be an opportunity to galvanize an organization's talent base and to move the organization forward in ways that would not be possible in times of stability.

- In chapter 4, we discuss how leaders can help their organizations move forward in the periods of uncertainty that are necessarily associated with disruption. We explore the nature of uncertainty and identify scenario planning as a tool that can help companies prepare for alternative futures by identifying certain types of strategic moves the organization can take amid uncertainty.

Part II addresses the technological infrastructure and associated business principles that have proven indispensable for helping companies navigate both chronic and acute disruption. Our focus is on the capabilities that these technologies enable for helping companies navigate disruption, not the technologies themselves.

- Chapter 5 sets the stage for this section on the necessary technological infrastructure by first identifying the organizational characteristics that this infrastructure is meant to enable—nimbleness, scalability, stability, and optionality. This chapter emphasizes the point that digital technologies alone cannot provide digital resilience, but it is the organizational capabilities that these technologies enable that do.

- Chapter 6 examines the first building block of enabling technology, cloud computing. Whether it be video conferencing, cloud-based productivity tools, or file-sharing infrastructure, cloud platforms are indispensable for adapting rapidly to new ways of working at scale while facilitating new interactions and keeping these interactions secure. By looking at examples, we will see how a cloud-based infrastructure has proven its value many times over in recent months.

- In chapter 7, we focus on how companies use data to adapt to disruption. Data enables companies to rapidly test hypotheses on how certain changes affect their workforce, providing valuable evidence for managerial decisions in the midst of disruption. Data serves as the foundation for machine learning, which is proving invaluable for helping companies adapt quickly to uncertain situations and will almost certainly be a disruptive force of its own in the coming years.

- Chapter 8 focuses on the need for cybersecurity amid disruption. With the rapid switch to remote work and the increase in activity on digital channels, it increases the importance that these channels are a safe, secure, and stable place for business to occur. Given the complex nature of cybersecurity in today's environment, this chapter is somewhat different from others in this section, more making the case that companies need to take cybersecurity very seriously rather than providing clear guidelines for how to do so.

Part III highlights the organizational building blocks that are essential for helping companies effectively adapt to disruption and builds on the prior section's focus on the technological building blocks.

- In chapter 9, we focus on what was likely the most high-profile and universal impact of the COVID disruption—the rapid shift from colocated to remote work. We examine how companies have navigated this shift and what they have learned from it. More importantly, we explore how companies can enable their workforce by intentionally redesigning the workplace as well as how and what work is done.

- Chapter 10 focuses on how those employees work together. Specifically, we note that digitally resilient companies tend to use cross-functional teams and manage those cross-functional teams differently by providing greater autonomy and evaluating those teams as a unit. Although we initially observed these characteristics with respect to companies dealing with chronic disruption, it has been associated with those dealing with acute disruption as well.

- In chapter 11, we discuss how the relationship between companies and its customers change as a result of disruption. For some, the relationship becomes stronger, because the company realizes it provides a service that is more essential in crisis. For others, it means that companies need to redefine the relationship with their customers when the very value companies provide customers is disrupted.

- Finally, we waited as long as possible to write the epilogue of this book. It has been both rewarding and challenging to be investigating and writing about a phenomenon with such future global implications. We return to our findings after several months to reflect on the lessons of this book as well as to identify what might have changed since we finalized the project for publication.

I Understanding and Leading amid Disruption

1 What the Pandemic Taught Us about Digital Disruption

> You should never view your challenges as a disadvantage. Instead, it's important for you to understand that your experience facing and overcoming adversity is actually one of your biggest advantages.
> —Michelle Obama

In the fall of 2019, most of us would have scoffed at anyone predicting that 2020 and 2021 would bring record-breaking unemployment levels, millions of shuttered businesses, the cancellation or postponement of sporting events ranging from high school baseball to the Tokyo Olympics, and tens of millions of infections and millions of deaths globally. Yet here we are, trying to make sense of the effects of the COVID pandemic that swept across the world.

Not all disruptions change the business environment to the same speed or degree. In discussing the COVID pandemic with senior executives, a recurring theme is how quickly and severely it affected organizations, forcing them to respond with extraordinary speed and vigor. "Slow, but steady" doesn't work given the dynamics of a pandemic. By contrast, prior waves of digital transformation included more opportunities for experimentation built around scalable, but carefully planned, pilots. In reflecting on why our insights about digital transformation were so applicable to the COVID crisis but also seemed so different, we borrowed terminology from healthcare to refer to the differences in the types of disruption.

In a way, the COVID-inspired wave of digital transformation resembles how physicians respond to *acute* medical conditions—rapid and dramatic interventions designed to stabilize the patient and lessen the immediate severity of the condition. The intervention is often not appropriate for the long term but is necessary to give the patient a respite and transition to longer-term solutions. For example, heart bypass surgery may be necessary to prevent an imminent cardiac condition that can then be treated with a daily medication.

The expectation with acute medical conditions is that they are temporary, although they sometimes recur or morph into persistent or *chronic* conditions (See table 1.1 for more detailed comparison). Chronic medical conditions cry out for sustained treatments that patients can tolerate over extended periods of time. Often patients with chronic conditions are never completely healed but can live comfortably with appropriate medical treatment and accommodations. Extending the previous metaphor of cardiac conditions, chronic treatments may involve long-term medication and lifestyle changes that prevent the chronic condition from becoming life-threatening.

While the COVID pandemic and response represents an acute disruption today, it may become a more chronic disruption over time. This increases the challenge for executives who are trying to lead their organizations through complex digital transformation journeys to determine what are the equivalent digital moves of open-heart surgery and what are the pivots more appropriate for responding to persistent

Table 1.1
Features of acute versus chronic disruption

	Acute	Chronic
Onset	Sudden/rapid and often severe onset	Slow-building and persistent
Symptoms	Obvious and attention-grabbing	Not always obvious; can be overlooked
Treatment	Requires rapid and typically dramatic response	Requires sustained treatment that must be tolerable over time
Duration	Is temporary, though it can develop into chronic conditions	Long-lasting and cannot simply be "cured"

and long-lived disruptions (e.g., the shift from on premises to cloud computing).

In the same way that a heart attack can serve as the proverbial wakeup call and impetus for lifestyle changes for those suffering from chronic heart disease, acute disruptions can serve as an opportunity for organizations to make fundamental shifts and to implement structures and practices that will enable them to thrive.

What's Unique about the COVID Disruption? Accelerating Transformation

If there is a unique aspect of the acute COVID disruption, it would almost certainly be the rapid shift to remote work, which we will discuss in chapter 9. This acute crisis struck at a time when most organizations were partially along their journey toward digital transformation. The nature of the disruption (e.g., social distancing) combined with organizations' plans to implement technology that helped address this problem accelerated the plans of digital transformation that companies had in place. The acute disruption also obliterated many of the barriers that had been preventing many companies' digital transformation plans, creating a unique opportunity to move those plans forward at an accelerated rate.

We have observed a dramatic uptick in the deployment of digital technologies that help reduce face-to-face interactions and safeguard customer and employee health and well-being. These digital technologies include consumer-facing applications such as grocery and food delivery services, business-to-business (B2B) e-commerce applications, and applications such as video conferencing that seem to have penetrated the consumer, business, and not-for-profit worlds. Searches for terms such as "contactless" increased sevenfold between November and late April, while the stocks of technology companies aligned with newfound customer health and safety concerns have skyrocketed. Many of our prescriptions developed in the context of chronic disruption have been applicable—perhaps even prescient—to the struggles many companies face in responding to the acute COVID disruption.

One company even tapped their lead digital transformation officer as the point person for managing the disruptions associated with the acute disruption. Sara Armbruster, vice president of strategy, research, and digital transformation (and recently tapped to be the next CEO) at the office furniture manufacturer Steelcase, described the rationale behind the decision.

> With digital transformation, you need to take a broad perspective and really understand the business and think about how the power of digital can be applied in different ways to accelerate the business. I think the same is true when you think about COVID. It's touched every single aspect of our business so decisions from health and safety, to finance, to operations, to legal and compliance, to innovation opportunities. I think that's why I was selected.

But we also find that acute disruption creates a stress test for the investments that organizations made in digital transformation and increases the importance of certain areas (e.g., technologies that make social distancing possible). We're reminded of the Warren Buffett quote, "When the tide goes out, you see who's been swimming naked," to describe companies that talked big about digital transformation but did little to accomplish it and are being exposed in the current circumstances.

In fact, many of our interview subjects spoke in surprisingly positive terms with respect to the impact of acute disruption on their business operations.

- Janet Pogue McLaurin, global workplace research leader at Gensler, observes that "the whole pandemic has been an accelerant for workplace issues that already existed pre-pandemic."
- Chris Dellarocas, associate provost for digital learning and innovation at Boston University, said, "These have been some of the best weeks of my career. It's been great seeing this collaboration. We haven't seen each other since March, but everyone has been 100% effective."
- Christine Halberstadt, vice president of strategic transformation at Freddie Mac, notes that "it's actually been a positive more than a negative, because it's brought an acceleration of how technology can move the business forward."

- Her counterpart at Freddie Mac, CIO Frank Nazzaro, echoes this sentiment: "The disruption didn't change anything with respect to our plans. We had a vision. It proved to management that the technology was capable, and it accelerated the implementation."

- Marc Schlichtner, principal key expert and founding member of T-Club at Siemens Healthineers, makes a similar observation, noting that "it's really incredible to see how a crisis actually brings us to the best as human beings. I was surprised by our management, which sometimes does not act in the speed I would desire, they saw that there is big opportunity and increased their speed and flexibility in decision-making."

- Brice Challamel, global transformation lead at Google Cloud, sums up this perspective from several respondents quite nicely, saying that "this situation is what my team and I have been preparing for our entire lives."

In fact, if there ends up being a "silver lining" to the acute COVID disruption, it is that in many companies, the pandemic is ushering in digital changes that are long overdue. We expect that the digital changes companies will be enacting in response to the pandemic will be the most significant since the Y2K crisis at the end of the last century, if not of all time. Although future disruptions will likely see some degree of digital response, it may not be as significant as the changes we see emerging from the current crisis.

Looking to Previous Disruptions for Guidance

We explicitly stated in the introduction that this book is not primarily about COVID. Instead, we hope it to be a book about disruption that leaders can look to for help navigating future disruptions, regardless of its nature. Many of our interviewees indicated that they look to lessons of previous disruptions for guidance on how to deal with the one they were experiencing. For instance, Matt Schuyler, chief administrative officer at Hilton, said that we've seen companies that have gone through precipitous declines in the past, like the dot-com bust in the early 2000s or the Great Recession in 2008, and we've tried to model our response off of some of those things. From these examples, we see that if you do the right thing and treat your guests differentially during times of crisis, they'll be loyal to you going forward. So, we went back to the last crisis and see what customer service companies did, and we just accelerated the responses we saw there.

Brian Baker, global people strategy business partner at the advertising conglomerate WPP, also noted reliance on lessons from previous disruptions to guide organizations through the acute COVID disruption: "I heard a leader say 'I have led through 9/11. I have led through the financial crisis. Because of those two experiences, I had that confidence and the courage to lean on empathy more. Those experiences made me ready for this one.'"

Even more routine disruptions can serve an opportunity to prepare for larger ones. Monty Hamilton, CEO of Rural Sourcing, a software development company that operates digital engineering teams in the United States, also noted the importance of relying on experience from previous disruptions: "My COO convinced me that we could go remote if needed, because we've done it in the past. If you're living in Augusta, Georgia during Masters week—that's disruption, and everyone works from home. If you're in Mobile, Alabama during Mardi Gras, everybody works from home. So, we've done it office by office, just never all at once and not for this long."

These examples underscore the value of understanding acute disruption as a guide for future disruption—the goal of this book. We hope that leaders can use our work to look back on the lessons forged during this time to guide decisions in future disruptions. Furthermore, we don't need to wait until the next disruption to extract the lessons we can learn from this one. Michael Aldridge, associate vice president of experience at Humana, notes that

I hope the lessons we learn in this moment permeate other areas of the business: they're not just for times of crisis. We've learned a lot of rich lessons, like our ability to act with speed and adapt to the changing needs of our customers. It is often easy to lose sight of what you just accomplished and move on to the next thing. I hope we internalize what we've learned from our response to be even better for what may lie ahead.

Indeed, highlighting these lessons companies can learn from acute disruption was one of our main inspirations for writing this book, extracting ones that can and should be applied months or even years in the future.

What Doesn't Change with Acute Disruption?

Are digital transformation principles and approaches developed in response to chronic disruption still applicable in the case of acute disruption? We think that they are—and particularly more so.

Cross-Functional Teaming and Agility

In our early research into how companies are responding to the acute COVID disruption, cross-functional teaming is becoming more essential as organizational silos are broken down to enable the company to respond to the common threat. Companies are forced into cross-functional teaming to provide a unified and rapid response across the organization, and those teams are increasingly being given greater autonomy out of sheer necessity. Lower-level employees are stepping up to take responsibility in ways that they may have been reluctant to do so previously. All these considerations enable the organization to be more agile. In his book *Team of Teams*,[1] General Stanley McCrystal notes that teams are important in times of information uncertainty, when it is not clear who needs to know what information and when they need to know it.

Interview subjects report additional reasons that cross-functional teams are important and, although intended to address the problem of chronic digital disruption, appear to apply to acute disruption as well. Christine Halberstadt of Freddie Mac notes that organizing differently allows the company to think differently. Dave Cotteleer, vice president of North America Sales for Harley Davidson, argues that since digital cuts across the entire organization, digital efforts need to be organized accordingly. Shamim Mohammad of CarMax, comments that cross-functional teaming allows greater experimentation since different experiments can be assigned to different teams.

Continual Learning

Many companies are ramping up webinars and other means of educating employees. Employees are turning to platforms such as Skillsoft to learn how to manage in this new virtual environment and are developing new skills such as data science and coding. Skillsoft has seen a threefold increase in consumption of its products since the onset of the pandemic. Further, by being thrust into new challenges and circumstances, many employees are learning on the job. And as organizations are forced to reinvent parts of their businesses, some of them are learning through experimentation and creating new knowledge stores.

Likewise, our recent interviews with senior executives suggest that companies that will be negatively impacted the most by acute disruption will be ones that simply try to "weather the storm" and go back to old established business models. In contrast, companies and individuals that will emerge from this crisis stronger are seeking to develop the digital capabilities necessary to thrive. Mark Onisk, chief content officer at Skillsoft, indicates that corporate learning initiatives are often becoming part of the glue that holds the social elements of the company together while employees work remotely.

Mission, Vision, and Values

A third commonality between how companies respond to acute versus chronic disruption is leadership's ability to clearly communicate a strategic vision for the company. We found a strong association with an organization's leadership and its digital maturity. In fact, the three most important facets of digital leadership we identified are having a transformative vision, being forward-looking, and being change-oriented. Deloitte's 2020 biennial report on technology leadership underscores this idea,[2] showing that 69 percent of survey respondents believe future technology leaders need to be change-oriented, have vision, be agile, and be innovative.

Having and clearly communicating a strong strategic vision for the company during both acute and chronic disruptions helps employees know how to respond when digital disruption radically changes the environment. When the old rules are no longer applicable—whether suspended due to acute disruption or because technology moves faster than regulation—it is important that employees know which overarching principles are guiding their work.

What Is Different When It Comes to Acute Disruption?

Despite these strong similarities in how companies respond to both acute and chronic disruptions, there are also important differences. While one may be tempted to assume that these differences make acute disruptions more difficult to address than chronic disruptions, that assumption would be a mistake.

The Knowing-Doing Gap

An example of a barrier to digital transformation that has been removed is the "knowing-doing gap," a term coined by Jeffrey Pfeffer and Robert Sutton more than twenty years ago.[3] In our 2016 survey, nearly everyone (87 percent of our survey respondents) knew that their industry was going to be disrupted to a moderate or great extent by digital technologies, yet few respondents (only 44 percent) felt that their companies were doing enough to respond to this disruption.[4]

Of course, that's one of the challenges of chronic conditions, particularly if they are asymptomatic in the short term. It is possible to simply ignore disruption and go on with business as usual, putting off the most difficult and disruptive changes to the future. Acute disruptions, however, are far more difficult to ignore. A surprising theme throughout our interviews is that many companies had plans—and even the technological infrastructure—in place to enact the types of changes that proved necessary; they had simply planned a much longer timeframe to enact them. Acute disruption collapses the knowing-doing gap because failure to act is simply not an option. For example, as various governments issue stay-at-home orders, companies have little choice but to figure out how to work remotely. Many companies already know what to do, and the acute crisis gives them the opportunity to do it.

Experimentation

Another major barrier we identify with respect to the chronic digital disruption is the willingness to experiment and take risks. In fact, the willingness to experiment is the single biggest barrier companies reported in responding to chronic digital disruption.

Acute disruption, however, seems to greatly reduce resistance to experimenting and taking risks. When one does not know what to do, one must figure it out, and experimentation is an extremely important tool for doing so. Ben Waber, president and cofounder of the workplace analytics company Humanyze, notes that he has been astonished at the extent that executives operating during the acute COVID disruption have been willing to try new things without necessarily being certain of the outcomes. Part of this is that many organizations were forced to act and take risks. In fact, it was clear that not doing anything or doing things "as usual" was a greater risk than taking action by trying something new and unproven. Adaptability as an organizational health indicator, particularly during rapid change, can help to support a successful workplace environment.

Acute disruption has also given greater free rein to innovators within companies. The digital innovation team at Siemens Healthineers reports digital innovation is more possible during acute disruption because the usual bureaucratic barriers that hinder innovation have been removed as a result of working from home—either through formal removal or by just being able to work around them more easily when not in the office. For example, the US government waived certain penalties instituted in the Health Insurance Portability and Accountability Act (HIPPA) for good faith use of telehealth during the coronavirus crisis to allow healthcare providers greater leeway to respond innovatively to the novel crisis.

Level of Uncertainty

Another difference is the magnitude of uncertainty that leaders have to deal with. In digital disruption, change can happen somewhat slowly and can be predicted with a certain degree of accuracy. As a result, we previously advocated for absurdly long-term strategic planning, suggesting managers plan for what is possible among several possible futures on a ten- to twenty-year timeframe. With dealing with acute disruption, we do not necessarily advocate such long-term strategic planning as we would continue to recommend for chronic disruption, because we are likely to experience a decade's worth of uncertainty in the next twelve to eighteen months.

A Good Response May Not Be Enough

Chronic disruption, by definition, typically provides leaders with time and opportunity to respond. They may not make the right choice regarding which actions to follow or may not act quickly enough, but they at least have time to do something. It is possible that acute disruption is so severe that the best actions a manager can take may still not be enough to avoid big losses.

For example, many companies in the hospitality and travel industries experienced a 90 percent reduction in demand over the course of a few days that would persist for months, if not longer. There are few, if any, strategic moves available to a manager that would allow them to completely offset such a disruption. Marriott did execute a number of fairly impressive steps to adapt and mitigate this disruption, such as ramping up their retail sales of Ritz Carlton merchandise to help people bring luxury to the home, pivoting their best-in-class call center to help New York State process unemployment claims, and launching direct-to-consumer communications to preserve vital customer relationships.

While these steps undoubtedly helped, the challenges faced across the hospitality industry and some other sectors are simply so massive that there is only so much even the best leaders can do. Although we will focus primarily in this book on the remarkable cases of companies adapting quickly to the pandemic as a testament to their impressive stores of resilience, we do not want to minimize the real challenges companies faced or imply that somehow if these leaders did something different that they could have avoided massive disruption. Many factors are simply outside the control of most managers, and all one can do is lead in the context of the environmental conditions that one finds oneself.

One Final Thought

Disruptions often result in loss. In the case of COVID, loss has ranged from job loss to the loss of opportunities to the loss of one's health to the loss of loved ones. Grief is often the emotional response to loss. As Scott Berinato writes in *Harvard Business Review*, in talking about the pandemic, "the loss of normalcy; the fear of economic toll; the loss of connection. This is hitting us and we're grieving. . . . Our primitive mind knows something bad is happening, but you can't see it. This breaks our sense of safety. We're feeling that loss of safety. . . . We are grieving on a micro and a macro level."[5]

Digital technology will never replace what we have lost from the acute COVID disruption and will never assuage all the grief that we have experienced. Further, we do not want to suggest that somehow the business opportunities presented by the acute COVID disruption are worth the cost in terms of lives and economic impact. But maybe a bit of technology, applied judiciously, may help recreate the connections that were severed when opportunities for physical adjacency and mobility disappeared. A friend of ours, in speaking about the role of technology is fond of saying "don't put the plumbing ahead of the poetry."

A resilience mindset is important for overcoming various impacts of disruption, whether they be personal and emotional or economic and strategic. In the next chapter, we talk about cultivating a digital resilience mindset that allows leaders and organizations to respond effectively to both acute and chronic disruption but particularly acute disruptions that afford leaders and organizations with less time to ponder their options and test their way into the appropriate solutions.

**How to Apply the Concepts from This Chapter: A Framework
for Understanding Disruption**

At its core, scenario planning aims to explore a range of possible "futures," including the challenges and opportunities they may present. These futures may result from chronic disruptions that will likely emerging from prevailing trends. Some of these may be certain (e.g., global warming, automation, 5G infrastructure) and others less so (e.g., geo-political instability, development of urban infrastructure). These futures may also result from predicting possible acute disruptions. For example, while no one accurately predicted the COVID disruption, Bill Gates cautioned generally about the future threat of an outbreak in 2015.[6] Scenario planning allows organizations to have a tangible point of reference from where to build strategies and plans, generate new ideas, and—most importantly—take action. We recommend organizations and executives answer the following questions to effectively conduct scenario planning along with the simple tools we provide in figure 1.1.

1. What *trends* (e.g., chronic disruptions)—macroeconomic, socioeco-nomic, technological—today are undeniably going to continue to affect markets, our customers, and our assets? See question 1 for an example of technology trends.

   What trends—specific to the industry we operate in—must we similarly consider? While we recommend that you choose the trends most relevant to your business and sector, we find that many trusted institutions[7] and consultancies provide a good starting point for both cross-functional[8] and industry-specific trends.[9]

   Note that trends are practically indisputable (e.g., rising popula-tions, severity of weather patterns, urbanization). Elements that are less certain are covered in question 2.

Question 1: Trends

Trend	Details
Automation	Increased deployment of hard and soft robots, providing cost advantages to early adopters and displacing certain job families
Aging workforce	The average age of the workforce is increasing, as birth rates decline, requiring some economies to rethink retirement ages, address underfunded public and private pensions, and make a significant investment in upskilling/reskilling
5G networks	Pervasive 5G infrastructure, increasing connectivity and unlocking new experiences, new technologies, and new business models
Data	Acceleration of data creation and capture, with increased strategic importance of data

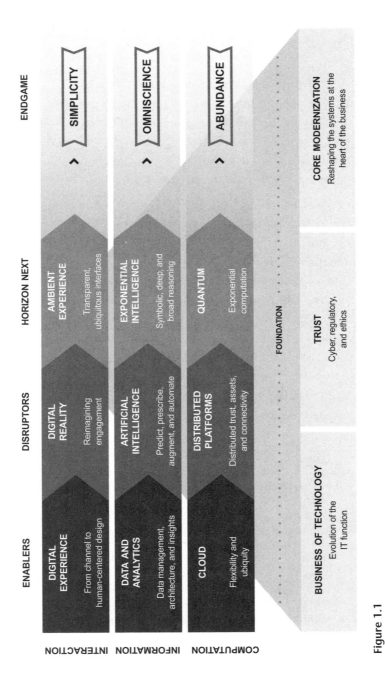

Figure 1.1
Technology trends for use in scenario planning

2. What *uncertainties* (e.g., potential chronic or acute disruptions)—macroeconomic, socioeconomic, technological—might affect markets, our customers, and our assets over the next five to ten years? The outcomes of uncertainty are, by definition, unclear and require us to imagine differing realities if they were to come to fruition; that is why we have two impacts associated with each. You could have even more! However, as a part of this question, also identify potential acute disruptions that might strike suddenly in a five- to ten-year timeframe.

3. What is the *likely outcome* of each trend and the *range of outcomes of each uncertainty* on our markets, competitors, and business? Include two opposing types of impact for uncertainties and be as specific as possible (see questions 2 and 3).

Questions 2 and 3: Uncertainties and likely outcomes (sample)

Uncertainty	Impact 1	Impact 2
Talent availability	Talent access is limited; changing workforce values and skillsets make it difficult to attract the right employees and to maintain engagement in the workplace.	Talent is readily accessible; companies address changing workforce values (e.g., curated work, new forms of engagement, meaning/purpose), and the workforce develops critical skills to address evolving requirements.
Healthcare[10]	The effects of COVID-19 shake society but, after a slow start, are met with an increasingly effective health system and political response. The virus is eradicated earlier than expected due to coordinated measures by global players to spread awareness and share best practices.	The COVID-19 pandemic becomes a prolonged crisis as waves of disease rock the globe for longer than anyone was prepared for. Mounting deaths, social unrest, and economic free fall become prominent. The "invisible enemy" is everywhere and paranoia grows.
Geopolitics	International relationships significantly deteriorate, especially between Western economies, China, and Russia, limiting cross-border collaboration and trade.	International relationships improve again, enabling greater cross-border collaboration and trade.

This exercise is only the first part of the scenario planning process that helps you think about chronic and acute disruptions you will face. We will continue the exercise in chapter 4, when we discuss what to do with it.

2 Beyond Surviving: Developing a Digital Resilience Mindset

Your ability to adapt to failure, and navigate your way out of it, absolutely 100 percent makes you who you are.
—Viola Davis

A fixed mindset is the belief that your intelligence and abilities are unchangeable. Its antithesis, a growth mindset, is the belief that your intelligence and abilities can change over time as a result of new experiences. The belief in a growth mindset is a belief that individuals can purposefully change their circumstances through hard work, practice, or training.

Carol Dweck, a psychologist at Stanford University, popularized these terms and described each mindset in detail in her best seller *Mindset: The New Psychology of Success*.[1] Dweck argues that how we think about our talents and abilities dramatically influences our success in school, work, sports, the arts, and almost every area of human endeavor. Dweck's underlying message will be familiar to parents who have read their children the classic bedtime story *The Little Engine That Could* about a small switch engine that makes its way up a high mountain through sheer grit and determination, naysayers notwithstanding.

A key takeaway from Dweck's research is that individuals with fixed mindsets are less likely to try something new to avoid the risk of failing. They stick with the known and shy away from the unknown. By contrast, individuals with growth mindsets make it their mission to learn new things, accepting setbacks and failures along the way.

Shortly after taking on the CEO role at Microsoft, Satya Nadella received a copy of Dweck's book from his wife, who had found it useful in helping her teach a daughter with learning disabilities. The impact of the book on Nadella was profound, and he attributes a growth mindset as key to the successful turnaround of Microsoft: "Dweck divides the world between learners and non-learners . . . demonstrating that a fixed mindset will limit you and a growth mindset can move you forward. Microsoft's culture change was 'the belief that everyone can grow and develop; potential is nurtured, not predetermined, and anyone can change their mindset.' They would shift from being 'know-it-alls' to 'learn-it-alls.'"[2]

Growth Mindset and Digital Transformation

This phenomenon of fixed versus growth mindset shows up repeatedly in our data on digital disruption (figure 2.1). Those who believe that their company is going to be in a worse position view their organization as being on the receiving end of digital disruption and the

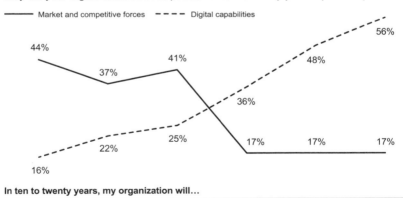

Figure 2.1
Growth mindset and digital disruption

associated market and competitive forces. When we talk with people at fixed mindset companies, they often assert that "we're just not a digitally native company" as if digital capabilities are inherent traits that some organizations just have and others just don't. They claim that their digital transformation efforts are stymied by the short-term thinking of Wall Street investors or competitive forces, all external factors.

In contrast, those who believe that their company will be in a stronger position as a result of digital disruption believe that it will be the result of their own hard work and persistence, coupled with a bit of good luck. They believe that their company will be capable of developing the skill sets and processes that enable them to thrive in a digital world. They have the capability to become what Nadella describes as "learn-it-alls."

Fans of Angela Duckworth's best seller, *Grit*,[3] and Malcom Gladwell's best seller, *Outliers*,[4] will see a connection between a belief in a growth mindset and the importance of effort. Judith Shulevitz summarizes the key insights in these books nicely: "If this book were a Power Point presentation . . . the best slide would be the two equations that offer a simple proof for why grit trumps talent: Talent × effort = skill. Skill × effort = achievement. In other words, 'Effort counts twice.'"[5]

The Digital Resilience Mindset: Innovating through Disruption

We have argued that developing a growth mindset is perhaps the single most important step toward Responding to chronic digital disruption at the employee, leadership, and organizational levels. While we still subscribe to this belief, we now recognize that a growth mindset assumes different forms when an organization is dealing with an acute versus a chronic disruption. It is this growth mindset that is the theoretical basis for the core thesis of the book—companies that emerge stronger will be those which innovate through the crisis.

With chronic disruptions, growth mindsets largely inspire companies to take on the types of changes that sustain their companies into the future. With acute disruptions, companies must quickly make the types of changes that allow them to immediately adapt to the environment that is rapidly thrust upon them. Companies with a fixed mindset orientation most likely regard the crisis as a challenge to be endured, with the expectation that they can eventually return to "business as usual." Companies with a growth mindset orientation may regard the crisis as unfortunate, taxing, and even tragic, but they see it as an opportunity to develop new types of organizational capabilities—capabilities that can help the organization navigate its way through the immediate crisis, as well as the "new normal" that follows.

Emma Lewis, general manager of shell polymers and global chemicals strategy at Shell, notes these differences in the oil and gas industry.

> I was looking at some analyst evaluation and it was the first time I'd seen an analyst talk about resilience. Some oil companies seem to be waiting for things to go back to normal. It appears they are planning to cut costs now and be ready to seize the day when it comes back. Others are more of the belief that it's triggered the energy transition happening faster. We are looking across all our businesses and at really decreasing some of our heritage activities.

Andy Ruben, founder and CEO of recommerce company Trove, speaks eloquently about the need and opportunity to innovate more aggressively amid disruption. He recollects his company's strategic moves in the weeks following the lockdown. He says, "In moments of incredible change, that's when the companies that are most nimble thrive. We asked what are 15 things that we've always known we've had to do, but never gotten a shot to do? We adopted the mindset that this is our moment; we can't miss the moment."

We use the term "the digital resilience mindset" to describe this way of thinking. Resilience is generally understood to refer to the ability of a material, a person, or an organization to Regroup quickly from difficulties and setbacks. A digital resilience mindset is one that combines this ability with a growth mindset and perseverance or toughness (what Duckworth would call "grit").

In many ways, our concept of digital resilience resembles what David A. Garvin, the late C. Roland Christensen professor of business administration at Harvard Business School, calls the "learning organization."[6] He defines a learning organization as "an organization skilled at creating, acquiring, and transferring knowledge, and at modifying its behavior to reflect new knowledge and insights." The primary difference between our concept of digital resilience and Garvin's learning organization is simple—the environment in which it operates. The nature of knowledge creation, acquisition, and transfer, as well as the need and ability to modify an organization's behavior, are fundamentally different today as a result of digital technologies than it was in the early 1990s when he introduced the concept. Organizations simply cannot learn or react quickly enough to compete without leveraging the capabilities of these digital tools. Even if the objectives of both are similar, the methods of accomplishing those objectives are so different that new competencies, new technologies, and even new terminologies are necessary.

Stages for Innovating through Disruption

Based on our research, we believe that when it comes to acute disruptions, organizations with digital resilience mindsets follow a three-step process—Respond, Regroup, Thrive—that starts by taking immediate and decisive action to deal with an existential threat, but recognizes the opportunity to build longer-term capabilities and become better prepared for future acute disruptions.

This perspective has been shared by several of our interview subjects. For example, Janet Pogue McLaurin of Gensler notes a similar response at her organization. She says, "We're thinking about this situation in three phases. What do we need to do to help our people and our clients while this pandemic is ongoing? How do we start to get people thinking differently? The final piece is reimagining the future and starting to pilot ideas."

Albert Baladi, CEO of premium spirits company Beam Suntory, identifies five priorities that have really guided their actions through the acute COVID disruption.

First, the most important thing was protecting our people whether they're in our operation facilities or in offices. Our people come first and are our North Star. Second, we secured our supply chain across our global markets in distilleries and operations facilities. Third, we continued to give back to society, which is absolutely central to our culture as a company. Fourth, we saw consumption starting to shift. We saw there's a new normal upon us here, and we need to be winning in the new normal. Fifth, we needed to anticipate future trends and use this opportunity to position ourselves to win in the long term.

The first three principles fit squarely into the Respond category, the fourth describes Regrouping, and the fifth principle captures the instinct to Thrive.

These stages are clearly distinct in higher education. In the spring of 2020, universities shut down and were forced to pivot nearly instantaneously to a complete online format. Over the summer, they Regrouped to figure out how to offer robust hybrid solutions, and in the fall, they executed a new plan in the disrupted environment. Chris Dellarocas of Boston University describes their efforts: "In early June, we put together the recovery effort, and this created a number of working groups, ranging from undergraduate to graduate programs to student life. We created a hybrid teaching model called "Learn from anywhere" to give flexibility to students that we are implementing in the fall."

Respond

It is difficult to cite a contemporary example to illustrate all aspects of our Respond, Regroup, Thrive framework since, as of this writing, the current global environment prevents companies from moving beyond the second stage. We thus turn to an example from an earlier disruption for illustration. In 2010, the eruption of the Icelandic volcano Eyjafjallajökull disrupted most of European air travel for a period of seven to ten days. KLM airline's response to that crisis demonstrates our Respond, Regroup, Thrive framework and ultimately helped the company to develop a "best-in-class" social media customer service organization.[7]

During the disruption, 107,000 flights were cancelled, stranding ten million passengers. Then VP of e-commerce, Martijn van der Zee, describes it as "a tsunami of communication," where everyone wanted to fly again but 95 percent of the available seats were already taken.

It was a bombardment of people trying to reach us through airports, call centers, and the Web. Our website can handle a lot, but it could not handle that number of people reaching out. That's when people started to communicate with us in volume through Facebook and Twitter, because it was the only way people could get through to us. We started to use Facebook and Twitter to give updates, which were very much appreciated. But, of course, when you give updates, people talk back. In the end, 160 volunteers helped the company Respond to the crisis—just communicating through social and working through our own system.

Much like companies' experience with acute disruption, KLM did not have a playbook in place for how to deliver customer service to Respond on social media. They adopted available tools on the fly to communicate with customers when the established channels could not keep up with demand. Yet, in doing so, they experienced firsthand the potential power of the platform for delivering a timely and efficient customer service experience at global scale. Van der Zee noted that "we never had a description about whether social is important to us and if we should engage or not. The crisis became our business case."

Over time, successful people and organizations get smarter about responding to acute disruptions. They develop playbooks that can be pulled out and adapted as circumstances warrant. That is why we participate in fire drills and listen to safety briefings prior to airplane departures. We recognize that we do not need to create a response from scratch when we can learn what did and did not work in the past. Military organizations have honed the art of extracting learnings from experiences, both good and bad, through after-action reports that analyze what happened and why. Similarly, the National Transportation Safety Board, an independent federal agency, investigates airline accidents and other major transportation incidents with the goal of identifying safety improvements that will make future occurrences less likely.

In the case of the acute COVID disruption, unfortunately, most companies in the West entered uncharted territory. Prior coronavirus epidemics had been limited to a handful of countries. Two characteristics of COVID—a high infection rate and the fact that asymptomatic individuals could spread the disease—made this epidemic dramatically different from prior ones. Yet, some Asian countries were far quicker to Respond to stop the spread of COVID because they had been exposed to similar epidemics such as SARS and MERS in previous decades. These countries had learned from their previous experiences.

While chronic disruptions elicit similar challenges with building new capabilities while continuing to keep the core business running, the difference here is the timeframe. Schools, businesses, and not-for-profits all needed to learn to work differently in a matter of days and weeks, not months and years. In times of acute disruption, the prevailing measure of success is not growth rates or profitability levels or market share but simply survival.

What we have learned from our research is that organizations with a growth mindset tend to be better prepared, more capable of swiftly executing, and more capable of shifting components of their business model in response to disruption. Growth mindset shifts from a focus on "what is the external environment doing to our business model?" to one of "what can we do to respond to this disruption?"

When the pandemic shut down many restaurants, Portillo's (a Chicago-based casual dining restaurant brand with sixty-two locations) quickly had to shift much of its traffic and loyal customers from dine-in to its drive-throughs, increasing its drive-through traffic by 40–60 percent. In addition, it had to adapt to a model where orders could be placed digitally and then picked up curbside. According to Nick Scarpino, Portillo's senior VP of marketing and off-premises dining, they shifted team members, who would normally be serving in their restaurants, to support call centers and delivery. The upside of this shift was that the restaurant didn't need to furlough any employees during the crisis; they repurposed them.

Whether dealing with an acute or chronic disruption, an organization's ability to respond with speed depends on how quickly it can detect a trigger that demands a response. In the case of acute disruptions, the need to respond is often glaringly obvious (or should be). But even so, early detection can make a difference. For example, before the United States went into lockdown in March 2020, CIO Kristin Darby of Envision

Healthcare was preparing to move the medical group's clinicians and clinical support teams to work remotely where possible. She said, "As a national medical group, we were actively monitoring the news coming out of Asia and Europe and planning how we would support our teams during what was going to be an unpredictable situation." Putting in measures to detect and sense developments and triggers early enables a more timely and effective response.

Being able to sense external developments is particularly important in chronic digital disruption, which is likely slow-building, triggered by a combination of factors, and can even be so subtle that it is undetectable—a bit like the proverbial story of the frog in the pot of warm water being slowly boiled as the water temperature is gradually increased. Leading management teams build routines to identify potential external and internal triggers and then sense for those triggers on a disciplined basis. Increasingly, there are risk-sensing tools powered by artificial intelligence (AI) designed to scan thousands of data sources in hundreds of countries and dozens of languages to deliver rapid response monitoring about events as they are occurring, through predictive intelligence and horizon scanning.[8]

Regroup

After the organization responds to the short-term crisis (e.g., ensuring that it can work safely and securely), the organization needs to adopt a slightly more forward-thinking mentality. Returning to our KLM example, they initially did exactly what we encourage companies *not* to do—settle back into the old ways.

Following the crisis, KLM simply abandoned the social media tools, returning to business as usual through websites and call centers. It was only several months after the disruption had passed that the organization realized that they had discovered a powerful new means of engaging in customer relations that should be formalized. Doing so, however, required that the company organize in a somewhat different way because—as we noted in chapter 1—the new capabilities cut across several different functions. Realizing the potential business value of these new tools, the CEO tasked Martijn van der Zee with figuring how to employ them in the context of normal customer service.

> The funny thing is that, after that kind of situation, you go back to normal. What that effectively means in a large company is that everybody goes back to their silo. Finally, at a certain moment about three months later, the CEO was so fed up that he said, "I don't care which department does it. We just have to organize it." And that's when we came up with a very simple social business strategy. It's really a one-pager. This moment is where we changed course compared to other companies.

It was only once the company had a chance to Regroup that they fully began to consider the potential business value of the social media platforms for customer service. So, they began to formalize the processes and began to hire employees to respond to customers on social media rather than relying on volunteers.

We call this second phase, Regroup. Unlike the frenzied and unpredictable Respond phase, Regroup is more settled, though still uncomfortable and uncertain. During this phase, organizations with a growth mindset can use it to shift from a defensive position and to a more offensive mode and a return to a market-facing posture. Leaders shift from managing the crisis and short-term thinking to developing mid- and long-term plans and charting out a path for the future, which could include anticipating how to reinvent the organization.

Regrouping does not imply going back to the old way but simply recognizing that once you have restored some sense of normalcy to business operations, you need to systematically consider how to take advantage of the opportunities presented by the disruption. Joseph Joseph, principal and global director of design technology at Gensler, notes that "during this time, there are a million ideas that come out, and you have to be very specific about what you take to innovate with and what you don't in order to spend your energy in proper places that return the most value to our clients and our business."

Portillo's, in addition to shifting its focus and staff to call centers and drive-throughs within a few months of the pandemic, launched an entire self-delivery program rather than purely relying on third parties. It enabled them to strengthen their brand and customer satisfaction by having delivery done by passionate and knowledgeable employees who go the extra step to take the time to pay attention to the details that the company prides itself on.

To help organizations through the Regroup path, leaders should ask themselves five fundamental questions to guide them through this phase and the next:

1. What is the new external business environment and how is it likely to evolve?
2. What does this environment mean for our industry?
3. What does recovery look like for us and what is our plan?
4. How do we activate the plan?
5. Of the changes that we implemented as part of Respond, which are still relevant (and can be a foundation for other enhancements) and which are no longer needed?

Thrive

The third phase of reacting to disruptions determines the winners and losers for both chronic and acute disruption. Once KLM had decided not to settle with returning to the "old way," they developed a strategy for implementing the lessons they learned during the crisis at scale. The company used the crisis as an opportunity to move the company aggressively into the digital world. Van der Zee concludes:

> The thing that seems most appreciated is still social customer service. We have a guarantee of Responding within one hour on Facebook and Twitter 24/7 in 10 languages. We are working on expanding that to 14 languages. So, you always get, in your own language, an answer within the hour. That's the guarantee. We are in contact with 30,000 people a week, which allows us to get the data from those people and to not only help them, but also to reach them in case of actions or other activities. We have now 130 people around the globe working on that. And it's so popular, the service, that it's still not enough. We intend to increase it even further because people simply

tell us, "this is, for us, a reason to buy, because it is so reassuring that we get an answer within 10 minutes without the pain of waiting on the telephone— just a rational answer."

Furthermore, the company began to discover new applications for its social media platform, such as a virtual lost and found. If a customer discovers that they have left a valuable item on a plane after they passed through security, they cannot go back and get it but KLM employees can. They began a program called "meet and seat," which allowed passengers in select markets to select who they would sit next to on a flight. The service was particularly popular in Brazil, which is surprising since they only had one flight to Brazil at the time, but the new offerings enabled them to expand their service. This ability to use the disruption as an opportunity to Thrive ultimately resulted in what one case study referred to as "best-in-class" social media customer service in the travel industry.[9]

The danger with acute disruption is the false belief that life will go back to a predisrupted normal. With certain types of acute disruption, such as a pandemic, it is easy to believe that once the storm passes and once a vaccine is developed, business will resume as usual. But during this acute disruption, employees and customers may have learned new habits, some of which they may be reluctant to shed. Understanding which new habits should be nurtured and which ones should not be is critical to effective action in this phase. Eric Ranta, director of North America cloud value advisors at Google Cloud, emphasizes the need to capitalize on the opportunities created by disruption.

> In the digitally resilient companies that I've worked with, you can almost sense a calm about them. Amidst the chaos, they know this is the world we're living in, and it will probably be living in it for some sort of period—whether it is measured in weeks, months, quarters, or years. Yet, they see opportunity everywhere. Companies that make bold bets in these types of times can actually have a very good result 3, 5, 10 years from now. The ones who are going to continue to invest are going to be the ones that are going to succeed in the future.

The Thrive phase is an opportunity to chart a new path, a new vision for the organization. It's a chance to implement changes that will build agility and resilience. This is where the lessons of chronic digital disruption are best applied. But it requires a commitment, tenacity, and growth mindset from leaders. As we noted in the last chapter, one of the silver linings of the pandemic is that it ushered in a wave of long-overdue transformation. The lesson here is that for many companies, it may not be too late to shift toward this growth mindset and begin to navigate your company toward this digital future. Brad Surak, former president of digital solutions at Hitachi Vantara, echoes this sentiment: "You're either going to thrive in this environment by being a first mover in it, or you're going to survive in this environment by moving, or you're going to die because you didn't do either."

Repeat

The lessons of 2020 may seem like a once-in-a-century anomaly. But the reality is that weathering disruption (both chronic and acute) is likely the "new normal" in itself. When we look back on the list of disruptions companies have endured over the past two decades we covered in the introduction, we are surprised at how extensive the list is—some caused by technology, others caused by politics, natural forces, and economic factors. In fact, even chronic disruption poses challenges to companies. Survey respondents indicated that the second biggest challenge report that their company faces with respect to chronic digital disruption is "ambiguity and constant change."

Acute disruptions may follow the linear path laid out in this model, as they may have more clearly defined before, during, and after periods. However, chronic disruptions may still involve these same steps but will occur in a more recursive cycle with blurrier lines between them and less severe and distinct Respond and Regroup phases. Nevertheless, if companies are dealing with multiple disruptions along different fronts, they may engage in different cycles at the same time. Managerially, it may be better to think about what disruptions you are dealing with and how and in what areas are you Responding, Regrouping, and Thriving. It may be most helpful to think about and manage disruption as a portfolio rather than as discrete events.

Leaders and organizations with a growth mindset will be better positioned to cope with disruption. Organizations filled with "learn-it-alls"
are much more likely to succeed than those filled with "know-it-alls."
Throw in a good measure of grit, and you have the basis for a digitally
resilient organization.

**How to Apply the Concepts from This Chapter: A Framework
for Assessing Your Capabilities**

A company's capabilities help determine how well it weathers and how
quickly it can move through the Respond, Regroup, and Thrive phases
of acute disruptions. Capabilities describe the set of objectives, processes, technologies, and talent that collectively generate value for an
organization and allow it to deliver its strategy (see figure 2.2).[10] Organizations with a growth mindset tend to be hyperfocused on identifying the capabilities that allow them to "play to win," recognizing that
strategies work only when the right capabilities are in place.

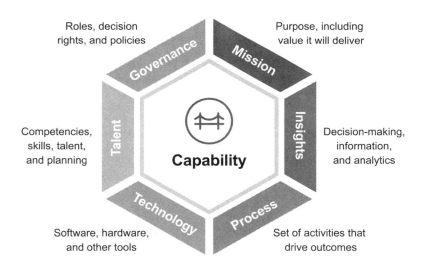

Figure 2.2
Components of capabilities

We sort capabilities into one of three categories: foundational, core, or strategic, based on the value added:

- **Foundational** capabilities are those that drive no differentiating value for organizations but are required to do business. For example, most companies consider employee onboarding a foundational capability. Foundational capabilities exist in service of core and strategic capabilities.

- **Core** capabilities are essential to the business you run and industry in which you operate. For example, compliance is a core capability for financial services businesses but would not be considered a significant differentiator or source of strategic advantage by those businesses.

- **Strategic** capabilities are those that drive a clear and significant competitive advantage; these allow organizations to command a premium in the marketplace (e.g., higher prices, shelf space, favorable deal terms). For example, product design is a strategic capability among consumer technology companies.

The same capability may be foundational to some companies and strategic to others. Take corporate development, for example. Companies that have a history of inorganic growth and success capturing synergies via integration, such as Cisco, may consider corporate development a strategic capability. On the other hand, companies that rarely make acquisitions may choose to outsource this capability to an investment bank or consultancy entirely and consider the capability foundational.

Companies often invest heavily in capabilities they believe to be strategic to their business, but they often underinvest in core and foundational capabilities. As a result, there is a high correlation between types of capabilities (segmentation) and their maturity. Companies that have a balanced investment strategy across the different types tend to fare better during acute downturns. They Respond quicker, Regroup more intelligently, and can "break into" the Thrive zone.

To help you assess your organization's capabilities, we introduce the capability matrix exercise. Here's how you use it. We will be referring back to this exercise in chapter 6 as well.

1. Brainstorm five to ten capabilities that are fine even if they are not mature, five to ten that need to keep pace with the market, and five to ten that will need to lead the market. Write each capability on a sticky note or index card, ideally using a different color for each category. Use the sample capabilities in table 2.1 to prime your thinking about these capabilities.

2. Reflect on the importance of these capabilities to customers, partners, and employees. Organize these capabilities into three groups based on their importance for your organization: foundational, core, or strategic.

3. Assess how you perform these capabilities relative to existing and potential competitors by grouping them into whether you are lagging, matching, or leading the market in these capabilities. Place the cards in the appropriate cell of the matrix based on the importance and maturity (see table 2.2).

Table 2.1

Sample capabilities across various functions

Finance and revenue management	Supply chain	Marketing	Sales	Human resources
Planning and forecasting	Warehouse and inventory management	Customer knowledge management	Business need identification	Employee life cycle management
Fixed asset management	Supplier management	PR and communications	Sales strategy and planning	Compensation and benefits
Management reporting	Product scheduling	Segmentation and targeting	Tech and thought leadership	Performance management
Accounting	Returns	Digital marketing	Quota setting	Onboarding
Revenue planning	Supply chain design	Brand management	Technical sales	Recruitment
Accounts receivable	Production planning	Product marketing	Services selling	Career and succession planning
Accounts payable	Demand planning	Localization	Sales enablement	Learning and development
	PO management	Media strategy	Solution selling	Diversity, inclusion, and equity

Table 2.2
The capabilities matrix

	Foundational	Core	Strategic
Leading	Maintain position	Maintain position	Maintain position
Matching	Maintain position	*Targeted investments*	**Invest rapidly**
Lagging	*Targeted investments*	**Invest rapidly**	**Invest rapidly**

Where you place the capability in the matrix implies how you should invest in it. For capabilities that are in lower right corner of the matrix, it is important that you invest rapidly or risk becoming obsolete. For capabilities in upper and left parts of the matrix, continue to maintain your position or even begin to claw-back investment, depending on where you sense the market heading. Most crucially, for capabilities in the middle and lower left quadrants, consider where you can make targeted investments. These capabilities are most likely to come in handy during acute disruptions.

You will see this capability-led approach come up a couple of different times in this book. As such, we recommend that you take the time—either alone or with trusted colleagues—to articulate and understand which capabilities are foundational, core, and strategic and equally, which ones are lagging, performing, and leading.

3 Digital Resilience Readiness: Leading through the Fog of War

I believe luck is preparation meeting opportunity. If you hadn't been prepared when the opportunity came along, you wouldn't have been "lucky."

—Oprah Winfrey

The "fog of war" describes the loss of certainty about one's own capabilities and the external environment under combat conditions. In such situations, the leadership skills of soldiers are tested as they adapt to unpredictable circumstances and are forced to operate with less than perfect information. This concept, if not the exact term, is attributed to Prussian general Carl von Clausewitz, whose magnum opus, *On War*, was published posthumously by his wife Marie von Brühl in 1832.

In an *Atlantic* article written relatively early in the acute COVID disruption (March 2020), Derek Thompson likens the impact of the pandemic to war: "What we're experiencing now is the fog of pandemic."[1] As Thompson suggests, disruptions like the pandemic bring much of the same sense of disorientation as war does.

Target CEO Brian Cornell describes how he has tried to respond to acute disruption: "It's safe to say that sitting here today, America is largely out of business, as many industries have idled capacity, as consumers are staying at home, working from home. Schools are closed, it's a very unique environment that none of us have seen before, and there is no playbook for how to react in this environment. We're writing the script each and every day."[2]

Given the large number of articles focusing on leading under uncertainty over the past thirty years, covering the 2001 World Trade Center attacks or the 2008 financial crisis, one might reasonably conclude that uncertainty is the norm rather than the exception for most industries. "Writing the script each and every day" may, in fact, be the only way to think about leadership during "the fog of disruption."

Disruption as a Crucible of Leadership

Having to reorient oneself due to disruption is not necessarily a bad thing. In their classic *Harvard Business Review* article "Crucibles of Leadership,"[3] written in the wake of the dot-com bust, Warren Bennis and Robert Thomas argue that one benefit of severe disruptions is that they often allow great leaders to emerge: "Extraordinary leaders find meaning in—and learn from—the most negative events. Like phoenixes rising from the ashes, they emerge from adversity, stronger, more confident in themselves and their purpose, and more committed to their work. Such transformative events are called crucibles—a severe test or trial. Crucibles are intense, often traumatic—and always unplanned." They note that crucibles "force leaders into deep self-reflection, where they examine their values, question their assumptions, and hone their judgement."

We hope that the current disruption spawns a new generation of outstanding leaders who can help their organizations weather the current crisis and build the skills and perspective to lead their organizations though future disruptions as well. In fact, one of the most rewarding aspects of writing this book was interviewing the leaders as they were helping their organizations navigate the crisis. Almost without exception, we emerged from virtually all of our interviews being inspired by the leadership on display by the people we interviewed. We had a front row seat for what we believe will be many great future leaders being forged, either personally or by others following in the example they set during this crucible.

Yet, leading through disruption and creating this environment of distributed leadership is very different from the type of command-and-control environment in more traditional organizations. Bennis and Thomas note that four skills are essential for leaders to learn from and lead through disruption. The first three traits—*engaging others in shared meaning, acting with integrity*, and *developing a distinctive and compelling voice*—are the subject of this chapter. The last trait they note that is critical for leading through the crucible of disruption is *adaptive capacity*—the ability to quickly grasp context and demonstrate hardiness—will be the focus of the next chapter.

Combining the insights of Bennis and Thomas with the insights derived from our interviews, we argue that leaders have three important tools at their disposal for leading through disruption—purpose, values, and mission:

- Purpose is about engaging others in a shared meaning around **why** your organization exists and why it does anything that it does.
- Values is about ensuring that all members of your organization act with integrity and communicating and embodying clear standards for **how** your organization does anything and everything.
- Mission involves leaders developing a distinctive and compelling voice that clearly identifies and communicates **what** your organization is seeking to do.

In the sections that follow, we discuss each of these tools in the context of digital transformation and digital resilience.

Purpose: The Why

In his book *Start with Why: How Great Leaders Inspire Everyone to Take Action*, Simon Sinek describes a golden circle that consists of three concentric circles with *why* at the core, *how* surrounding it, and *what* in the outer layer. He describes how many organizations and leaders make the mistake of approaching the circle from the outside in, starting with *what* they want to accomplish, instead of *why*. Answering the question of *why* an organization does what it does, what is its purpose, its raison d'être, was important before the challenges of 2020, but in times of crisis, it becomes a north star, a beacon of light through the dense fog. It's hard to know what your mission or goals should be if you are unclear about why you exist.

On August 19, 2019, 181 CEOs from top organizations signed an open letter from Business Roundtable entitled "Statement on the Purpose of an Organization."[4] It was a formal recognition that businesses need a purpose beyond making money. Having a purpose is more than just having a goal or a mission; it must go beyond a company's products or services and is why a company exists—beyond economic exchange. It must be aspirational and inspirational. Purpose was a major factor in many of our interviews, whether it be from the healthcare companies trying to heal the sick, spirits producers creating hand sanitizer, financial services firms trying to provide stability and security to their customers, to companies in the hospitality space trying to take care of their employees amid an unprecedented drop in demand.

For many companies, the purpose was clear. For example, the team of 25,000 physicians and advanced practice providers at Envision Healthcare, which cares for more than 32 million patients a year across the United States, knew it was going to be on the front lines of the fight against COVID-19. Kristin Darby of the medical group noted, "We rallied around our work and the public health measures because we knew we had to do everything we could to slow the spread of COVID-19 and protect our clinicians so they could continue to care for patients. Both clinicians and our clinical support staff—like myself—felt a renewed sense of purpose."

Others sought to generate that purpose when it wasn't immediately clear. Eric Schuetzler, VP of global research and development at Beam Suntory, described how the shift to producing hand sanitizer, which we will describe in greater depth in chapter 11, helped provide a sense of purpose. He said, "You may be surprised how motivating it was for our team members. Let's be honest, we're not going to work every day and creating the next vaccine or performing the next heart surgery. It was something people could rally around and feel good that we were doing something to help the public good, by supporting our frontline medical workers."

Purpose-driven organizations stand for and rally around something bigger than merely selling their products and services. Purpose helps identify the group's goals, mission, and priorities. It helps people focus on what is important within the organization and also shifts people's orientation to outcomes rather than just activities.

In the context of digital resilience, this means that digital transformations undertaken in the midst of chronic or acute disruptions are more likely to be successful when leaders and employees are able to see transformation as more than a struggle for survival. Studies suggest that purpose-driven companies outperform the market fifteen to one, experience higher market share gains, and grow, on average, three times faster than their competition.[5] Rajeev Ronanki, senior VP and chief digital officer at the health benefits company Anthem, Inc., also noted the importance of purpose.

> A purpose-driven company and purpose-driven teams are central to making this work. As we rallied around the common purpose of addressing the COVID crisis, it's shown that the company can change at a pretty fast clip in a very meaningful way. It was really our purpose and beliefs that played a central role in improving the health of our consumers and our communities and amplifying that purpose. It helped us tremendously.

In times of disruption, purpose becomes even more important. Disruption creates considerable anxiety, fear, and grief among employees when uncertainty upends familiar day-to-day routines. A clear sense of purpose provides a rallying cry around which employees can gather and fosters an esprit de corps that serves as an invaluable resource for motivating work amid the disruption. The chief digital officer of a large, global insurer noted the importance of purpose for his company amid acute disruption.

> The company has been really disciplined and thoughtful around articulating our purpose and bringing it to life. And that has coincided with a moment to truly be there for our customers and help them rebuild that confidence in a really uncertain world. We exist for when people need support, protection, and stability. Purpose became the reason why people are innovating at this phenomenal speed. We had this before COVID, but what a perfect moment for us to live our purpose.

Research in psychology shows that people are more likely able to persist through challenges when they are driven by intrinsic motivators rather than extrinsic ones. Among the most powerful intrinsic motivators are feelings of autonomy, growth, and meaning.[6] Finding meaning or purpose in what you do, having a reason for getting up each morning, is perhaps one of the most powerful driving motivators, particularly in times of changes, difficulty, or uncertainty.

In chapter 2, we discussed the importance of having a growth mindset—a concept that starts at the personal, individual level but can be applied at the organizational level, as reflected in the culture, beliefs, and behaviors of the group. Likewise, purpose is something that each individual must find on his or her own; but when organizations define and signal their purpose, they create connections with individuals and communities that rally around shared beliefs and motivations.

Companies that are driven by a clear purpose can more effectively create strong connections to both their customers and their employees. As Sinek points out in *Start with Why*, "people don't buy what you do; they buy why you do it. And what you do simply proves what you believe." Marketers will tell you that brand loyalty is built on loyalty and trust, emotional factors, not solely based on facts and features.[7] Starting with a sense of purpose or a "why" creates neurological connection with the emotional part of the brain. In the same way that an organization's purpose garners loyalty and trust in customers, it does so in employees. It inspires employees to be engaged, to go the extra mile, and to power through to find solutions to difficult problems.

Purpose can help organizations rally during times of crisis, but it is also an important part of setting a longer-term path forward. Brice Challamel of Google Cloud starts all of his conversations with customers by getting them to articulate the purpose of their cloud transformation—not the technical capabilities a cloud platform can bring them but what the company dedicates itself to achieve for people, as an organization. He says the following:

> We need to be the master storytellers of change. It is the core of my job to receive someone who comes with a stack of unfathomable problems in their

minds which they can't even exactly define, and from there lead them to build together the vision for an ideal future. They are there to serve a purpose, even if they haven't reflected on it for some time, or simply don't know what it is, yet. So, we help identify the "why" of that organization and an ideal future state, and lay out the roadmap of how to get there.

Building your organization on a clear and strong understanding of *why* you do the things you do can serve as a strong brand differentiator during normal times but can really act as a guiding light in times of crisis.

Values: The How

How you do anything is how you do everything. It's an adage that speaks to how the way you handle any task reflects your beliefs and how you might handle any future task. If purpose is *why* companies do what they do, then values determine *how* they do what they do, operating as guideposts for behaviors and actions. It may seem counterintuitive to talk about *how* you do something before you talk about *what* you're actually trying to do. But if we frame this less about the steps needed to complete a task or activity and more about the belief systems that guide the *way* things are done, it may make more sense. Before approaching any mission, it's important to know who you are and what you believe in.

Values are what an organization believes are important about the way it operates and works. When disruption threatens normal operations, it can become unclear what are acceptable or unacceptable ways to address unexpected challenges and problems. Values help employees understand what the company believes is important about the way it works and communicates those values to employees.

Brian King, global officer of digital, distribution, revenue strategy and global sales at Marriott International, noted the importance of values for helping Marriott weather an unprecedented disruption in the hospitality industry.

> I think one of the reasons why we're going to weather this storm—it comes down to a value system. And it's a value system that's been in place for 93 years, which can be summed up in one simple phrase, "Take great care of your

associates, they'll take great care of the guests, and the guests will come back time and time again. -JW Marriott." You must have a belief system that is rock solid, and if you do, your team will get through the most difficult of times.

Marriott then uses this value system to determine their priorities and actions. King talks about how they've had to draw on this value system in times of tough decisions.

> Our value system has always been strong, but in times like this it becomes even stronger. We have associates who are unfortunately furloughed. It's heartbreaking. When we were forced to make this awful business decision, we don't send some generic form letter in the mail telling them their fate. Every leader was helping call thousands of associates. When I made those calls, the one thing that gets to me every time is these folks say, "I understand." They know the entire travel industry is struggling, they read the paper. "We just need you to be successful, we want Marriott to make it, because we want to come back one day." That doesn't usually happen when you are forced to lay people off. And that's because there's this value system—a belief system—that they're sincerely cared for.

Will these values, indeed, be enough for Marriott to weather this disruption? Only time will tell, but it certainly helps ensure that its employees are all pulling in the right direction and giving their all to make it happen.

Values also help employees operate in an uncertain environment. Companies that are forced to adapt significantly in the face of disruption are faced with a greater likelihood of making mistakes—and not just trivial ones. For this reason, a strong culture of integrity is a critical companion for teams at CarMax. Their mission is "to drive integrity by being honest and transparent in every interaction." Shamim Mohammad, CITO at the company, emphasizes that "integrity is core to everything at CarMax." As you give teams more autonomy, it is imperative that they also understand the company's values to guide that autonomy.

One early Uber employee we interviewed notes that the problems they faced during their period of hypergrowth—a different type of acute disruption—stemmed from a lack of clear values that resulted in the widely publicized ethical problems at the company. As Uber grew rapidly, many employees were promoted quickly, and there was no clear directive from the organization's leadership about how they should execute on their mission (and how they shouldn't). The lack of clear guidance about the company's values led some managers who were experiencing greater than expected autonomy to allow a toxic culture to develop in pockets of the company. An autonomous culture and entrepreneurial spirit are good enough values (i.e., how), but they need to be checked against other values such as how people should or should not be treated.

Incoming Uber CEO Dara Khosrowshahi spent much of the first two months in his position learning what worked at the company and what didn't, using that experience to develop a set of cultural values that would guide the company going forward. Khosrowshahi described the process for developing these values.

> Our values define who we are and how we work, but I had heard from many employees that some of them simply didn't represent the kind of company we want to be. I feel strongly that culture needs to be written from the bottom up. A culture that's pushed from the top down doesn't work, because people don't believe in it. So instead of penning new values in a closed room, we asked our employees for their ideas. There were some common themes: many people liked how the spirit of the previous values encouraged problem-solving and speed, but they wanted to see more around inclusion, teamwork and collaboration. They also wanted to make clear that we will put integrity at the core of all our decisions, and that we're unafraid to admit mistakes when they happen. We're calling these cultural norms, rather than values, because we fully expect them to evolve as Uber continues to grow.[8]

Values that are abandoned in times of difficulty aren't really values at all. For example, when asked about what concerns he had in the wake of the disruption, Brian Baker of WPP said that he worried at first what the crisis would do to their efforts at cultivating diversity in the organization. However, he saw the organization, and leaders, stay committed to the diversity and inclusion commitments made publicly by CEO Mark Read. This focus on preserving diversity amid a crisis demonstrates that it's a real value, not simply window dressing. A crisis also helps highlight what your organization's values really are.

Mission: The What

While it seems odd to cover the crucial "what" last, it is only by under-standing who you are as an organization (why you exist and what you believe in) that you can be clear on what your mission is, what your goals are, and what you're trying to accomplish. For example, *Mission Impossible*'s Ethan Hunt, portrayed in the series of movies by Tom Cruise, was clear on why the Impossible Missions Force exists (to get rid of bad guys and save the world) as well as guidelines for what was acceptable and unacceptable means of accomplishing his goals. While purpose and values remained consistent, the specific mission could change from movie to movie.

In times of crisis, it's critical to be clear on what your goals are. A mis-sion answers the question about *what* the organization wants to accom-plish. When disruption affects normal operations and decision processes, a clear sense of the organization's mission is essential when employees encounter unexpected situations and uncertainty. In many business organizations, the mission is a meaningless set of words printed on a plaque hung in the lobby and promptly forgotten, not a self-destructing message delivered in a dramatic fashion.

In the organizations we interviewed, however, mission became an important aspect of navigating the disruption. When normal work routines are disrupted and one cannot perform one's job in the usual way, the mission helps employees figure out how they can and should add value to the organization. For example, David Quigley, provost and dean of faculties at Boston College, described the role of mission in their decision to return to in-person learning in the fall of 2020: "Probably this year, more than any other time in my career, mission has driven so much of what we're doing. The idea of leaning into the possibilities of hope in a darker moment for the region, for the nation, and the world was an important part of it, and it helped guide our decision-making."

Mission can also highlight tensions for the organizations that need to be navigated and managed. Michael Lochhead, executive VP at Bos-ton College, also discussed the challenges associated with mission.

Mission was a double-edged sword in some ways here, though. Where the mission factored was trying to figure out how do you bring back the students in person, who are relatively low-risk for significant COVID impacts, while protecting all members of the community. Mission-wise, we were driven by a sense of care for all members of the community. It was one of those great tensions that guided our reopening process and decision-making.

A good example of the importance of mission is the military concept of commander's intent. A long-established military aphorism is that no plan survives first contact with the enemy. In chaotic situations, the mission (often embodied in a commander's intent) is essential in guiding an organization's response to uncertainty. In *Harvard Business Review*, Chad Storlie writes, "Commander's Intent is the description and definition of what a successful mission will look like. It empowers initiative, improvisation, and adaptation by providing guidance of what a successful conclusion looks like. Commander's Intent is vital in chaotic, demanding, and dynamic environments."[9] In his book *Team of Teams*, General Stanley McChrystal emphasized the importance of ensuring that members of the organization have a clear understanding of this mission. Drawing from his experience seeking to combat the Islamic State, he learned that having a big-picture sense of the organization's overall mission was essential for enabling members of an organization to make decisions and actions in an uncertain and complex environment.

Digitally resilient organizations, like military units, operate with the equivalent of a commander's intent to provide employees with the freedom (and boundaries) to make decisions amid an uncertain environment. With a clearly communicated mission, teams can understand what the organization is trying to achieve and can focus on creatively and nimbly determining how to execute pivots and adaptations while staying aligned to the mission and vision. Several of our interview respondents referred to the importance of mission in guiding their company through acute disruption. For example, Noah Glass, founder and CEO of the digital restaurant platform Olo, referred to the importance of their mission working with restaurants.

Our mission statement hasn't changed since we wrote it in 2013—"helping restaurants to better meet the needs of the on-demand consumer." Now, every consumer has now been turned into an on-demand consumer. I was able to talk about our mission and how important what we were doing was in this moment of crisis in the world and in the industry.

Brad Keller, director of workplace strategy at health insurer Humana, noted the importance of mission in the healthcare industry. He said, "Humana's mission is simply to help people achieve their best health. With a mission like that, during a time like this, everyone is empowered to improve the health outcomes of our members and to make the process simpler." Yet, Humana realized that truly executing on that mission in a disrupted environment required new and different actions, as Michael Aldridge explained:

We were hearing in our phone calls with customers that food insecurity was beginning to become a real concern with our members. They weren't calling with that purpose; it just came up in conversation. Food provision has never been a service that we provided at this scale, was never one our customers would have expected us to provide, and—frankly—it was not one we were skilled to provide. Yet, we realized that under these new conditions it was absolutely part of our mission, and we stepped in to help where we could. Our small team basically had to learn about food distribution in less than 24 hours, and we've now provided over 1M meals to our members.

Communicating the Mission

Yet, the importance of mission must be supported by a leader's ability to communicate that mission and keep the organization reminded of it. Many of our respondents indicated an increase in communication from their leaders at all levels. Albert Baladi, CEO of Beam Suntory, notes the importance of communicating the mission.

We knew that we couldn't manage all the decisions in a crisis mode from the center, so we saw the need to provide guidelines to our global teams. We've really doubled down on communication like we've never done before. Communication is obviously very important anytime, but especially in times of crisis, I don't think one can overcommunicate. Personally, I never thought that I would communicate as much as I have. By doing so, I think we were able to be very agile through real time sharing of information with the organization.

Yet, the biggest difference was not necessarily that the leaders were communicating more but that they were communicating differently. Virtually all of our respondents note that the tone and character of their organizations changed considerably amid acute disruption. Leaders became more willing to express uncertainty and fears to their employees. Although many of our respondents noted this increase in authenticity, Brad Keller of Humana addressed it most directly.

> What I've seen is a new level of candid authentic leadership. Our CEO has found this new voice of communication with our employees. We get this Monday letter from him that we all look forward to. It's the most candid letter you've ever seen from a CEO. It's a very authentic conversation and even sometimes showing vulnerabilities. That leadership has really resonated with me. It feels good to not feel so corporate.

Emma Lewis of Shell reflected on her attempts to be more transparent and authentic as a leader in the midst of the fog of disruption.

> In a crisis you try to lead with authenticity, great transparency, and with a real learner mindset. For example, the first call I had with my team after the whole George Floyd incident. I mean, I'm European. I'm sitting in the US, and I don't understand the dynamics, because I didn't grow up here. I also have a very personal view on it. So, I just was very honest with them. I said, "I don't know what the right thing to say is, is I'm just going to tell you how I feel." And I think it's just being willing to be vulnerable and admit that you're struggling, too.

That sense of authenticity, sincerity, even vulnerability is critical in strengthening trust and human connection, particularly in time of crisis and uncertainty.

Understanding how to communicate during a crisis is also important. For example, after the 2011 Japanese earthquake, Cisco had difficulty communicating with its 1,400 employees who were affected, and it took them a week to even confirm that everyone was safe.[10] This led Cisco to streamline their communication systems, which paid off during the 2016 terrorist attack in Brussels, enabling them to locate and communicate with all affected employees in less than twenty-four hours.

How to Apply the Concepts from This Chapter: Questions Guiding Purpose, Values, and Mission

Admittedly, forming a purpose, establishing a set of shared values, and defining a mission is easier said than done. Therefore, we are going to spend a little time thinking about how to articulate these in more concrete terms.

First, pick an organization, division, or a group that you belong to. Brainstorm answers to the following questions:

- Whom do you serve?
- Who benefits from your organization's existence?
- How do they benefit?
- What are the factors that inspire trust among your customers?
- Why do others permit you to exist?
- How does this align with what you want to be known for?

Write down your organization's purpose. Try to think as broadly as possible, but don't be afraid to go narrow if you picked a function or line of business. Your organization may already have a defined and published purpose statement, so feel free to use this as inspiration but do not feel constrained by it.

If you are short of ideas, some examples include the following:

- nourishing families with healthy, sustainable, and affordable food
- moving people and products through the air safely and reliably
- providing the necessary infrastructure to allow people to communicate and connect anywhere, any time
- ensuring that heads of household can affordably manage the risks of life and provide for their families no matter what

Now, let's unpack the values and behaviors of your organization— the "how." Like purpose statements, your organization may have these explicitly defined, but you should not take these at face value—often, we see a disconnect between an organization's stated values and how they behave. Think of a time, at your current organization or a previous one, where you noticed such a disconnect:

- What was it about that organization that you did not like?
- Were the leaders taking credit for things they did not accomplish or shifting blame?
- Were team members driven primarily by financial incentives?
- Were they encouraged to get the job done at all costs, even if it meant cutting corners?
- How did you feel? How did your colleagues feel?
- What happened when things shifted in the market?
- Did people leave or stay?

Think about the things that are important to you, your peers, your team members, and your leadership now. What is important to you? Your team? Your c-suite? Your board? Your employees? Your customers? Do they align? Triangulating viewpoints help to identify areas of alignment and gaps that need further work to create alignment.

Now write down what is most important to you and your team members gauging whether your personal values are shared. It is important to note that the embodiment of the values are your people and the decisions they make—in the short term it may cause hardship—so it is important that the values your organization defines must be easy to understand and shared.

If you are short of ideas, some examples include the following:

1. Take care of each other, every time, no matter what.
2. Always ask why.
3. Respect every interaction, every customer, every time.
4. Do the right thing.
5. Be as honest and transparent as possible.

Last, let's address the mission—the "what." You probably have a pretty good idea of what your organization is focused on—if you are like most organizations, it is the same old suspects—cash flow, market share, customer experiences, cost reduction, market expansion, talent, and likely another objective related to tangential growth opportunities. But what happens if you reframe this discussion in the context of your purpose and values? The notion of your mission may then look a little different. Write it down.

If you are short of ideas, some examples include the following:

1. Leverage the latest technologies in everything that we do.
2. Build sustainable businesses with durable competitive advantage.
3. Take calculated risks, but limit the costs of failure.
4. Create multigenerational brand value and loyalty.

Understanding the purpose, values, and mission, you can now set the conditions and parameters by which all decisions are measured and evaluated. This is not an exercise that you will want to perform alone—after all, this isn't just about you. Get out there and start talking to colleagues at all levels of the organization about what they think the purpose, values, and mission are. Do you notice any variations in perspective? Maybe the values and purpose that you thought was apparently clear to everyone else isn't. Maybe what is published is not truly reflective of reality. Now is the time to revisit, refine, and reshape based on your aspirations and value system.

4 Digital Resilience Readiness: Making Strategic Decisions in the Face of Uncertainty

Our anxiety does not come from thinking about the future, but from wanting to control it.
—Kahlil Gibran

In the midst of responding to disruption, it is incredibly easy to lose one's way. While disruption can fundamentally threaten a company's business model and practices, it can also present opportunities for unexpected pivots and adaptations.

Pompanoosuc Mills, a small manufacturer and retailer of carefully crafted home and office furniture, was founded in 1973 by Dwight Sargent, who still serves as its president. Based in Thetford, Vermont, the company was hard hit by the acute COVID disruption. Sargent laid off almost his entire 115-person workforce. But then he embraced digital marketing and online retailing, including implementing showroom tours on the company's website. Business started to recover, allowing Pompanoosuc Mills to "bring back workers that were previously laid-off and keep their eight retail showrooms active when social distancing restrictions were enforced." With online marketing, immediate delivery, and cut-rate prices, Pompanoosuc has received enough online orders to be able to rehire its workers and pay them bonuses."[1] Sargent hopes to outlast the pandemic but expects a very different future if he does: "Some of these things are exciting. We are so far behind."[2]

In the case of Pompanoosuc Mills, acute disruption forced a leader to make what was probably a long-overdue decision on moving forward with digital transformation. In other cases, acute disruption has required leaders to confront tough decisions about whether to cut product lines or lay off employees or abandon unprofitable markets and customers or refocus on new offerings more in tune with the current times (the Pompanoosuc Mills landing page alludes to the work-from-home trend, with the words "Work from Homework"). Time will tell whether the leaders reacted quickly enough or in the right manner.

During the pandemic, we also witnessed enormous growth surges experienced by companies such as Amazon, UPS, and Zoom resulting from the move to deliveries by mail, video conferencing, online shopping, and cloud computing. In some cases, these demand spikes created significant operational challenges that threatened to overwhelm these companies and alienate both loyal customers and desperate new customers looking for alternative suppliers.

The winners were the ones that retooled themselves quickly. Amazon's Day One blog started to include "updates on how we're responding to the crisis" that highlight "what we are doing for employees," "what we are doing for customers," and "what we are doing for communities around the globe." The July 28, 2020, update, for example, describes actions taken relative to employee bonuses, new safety precautions, unattended delivery options, and donations to underserved communities.[3]

Where to Play? How to Win?

The danger of the success stories common in the business literature is that they tend to tell the success stories in hindsight, after the manager's decisions have been made and paid off successfully. They don't tend to dwell on leadership mistakes, even if there are valuable lessons to learn from them. Yet, leaders need to make decisions during great uncertainty. Only time will tell whether each company made the right decisions and executed these decisions in the right way. But in the moment, the question is, How should leaders and organizations imagine choices and choose which paths among those options to pursue?

And what criteria should guide leaders in making these decisions? That is the focus of this chapter.

In each case, the companies faced a tough set of choices. We group the questions from the preceding paragraph into two general categories:

1. *Where to play?* Should they adapt their existing business model and operations to the new disrupted environment? Should they pivot their business model and operations to capitalize on a new opportunity made possible in the current business environment? Or should they simply stay the course in terms of products and services offered and customer segments served?

2. *How to win?* Once you decide whether to adapt your business model and operations, the next set of questions involves digital technologies employed to create loyalty and differentiation. For example, what capabilities and management systems are required to move into this new space, what are the components of the operating model, and where should there be new or different channels (e.g., virtual in lieu of face-to-face)?

Leaders in disruptive environments need a vision of what can be and the ability to help bring it about. But how do leaders get such a vision in the midst of uncertainty or enough of a vision to be able to act decisively? Figure 4.1 shows these questions and associated considerations to answer and enact them.

Four Levels of Uncertainty

In our conversations with chief strategy officers at a number of companies, many advocated that it would be much better to just wrestle with and understand the nature of the uncertainty rather than try to predict the future. We suspect that one leadership characteristic that will help distinguish the winners from the losers will be those that can deal with and act within ambiguity, compared to those trying to create a false sense of certainty. Michael Lochhead of Boston College emphasized the importance of teaching students how to deal with uncertainty. He said,

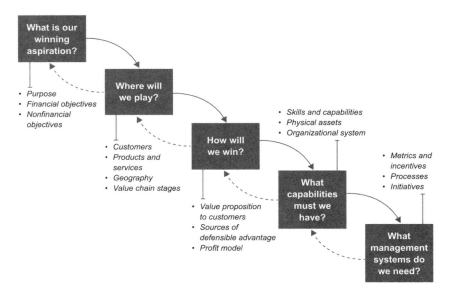

Figure 4.1
Questions to ask about where to play and how to win

"Our job as educators is to help prepare students and, inasmuch, we need to teach them resiliency. It's about helping students prepare for a world of uncertainty and to help them develop the skill set to take their education and apply that to a range of circumstances or outcomes that they can't predict or necessarily anticipate."

Yet not all types of uncertainty are created equal. In an article written for *Harvard Business Review* almost twenty-five years ago, Hugh Courtney and colleagues differentiate among four distinct levels of business uncertainty:[4]

- **Level 1: A clear-enough future**. We might call this first level of uncertainty "business as usual." The business environment always contains a certain degree of uncertainty, but it can be managed and planned for with a reasonable degree of confidence. Existing strategic plans will still apply to most possible outcomes.

- **Level 2: Alternative futures**. This level of uncertainty may necessitate a limited degree of shift in strategic action. The number of possible outcomes is relatively limited, and businesses can develop a course of action for each scenario. A company or industry facing a pending regulatory decision is a classic example of this level of uncertainty with two possible outcomes—the regulation passes or it doesn't.

- **Level 3: A range of possible futures.** The future could involve a range of possible outcomes, and it is difficult to know where the eventual outcome will fall in that range. Scenario planning is a particularly helpful tool for this level of uncertainty. The company plans for several possible outcomes and then adapts those plans as needed to address the reality as it emerges.

- **Level 4: True ambiguity.** Level 4 uncertainty arises when companies face entirely new business challenges that involve multiple possible uncertainties, making it difficult to engage in strategic planning. True ambiguity is difficult to plan for because they can involve unexpected patterns and exponential change. The key here is to take actions that can help reduce uncertainty and control what you can control until the level of uncertainty is reduced.

The initial phase of the acute COVID disruption clearly represented Level 4 uncertainty. Andy Ruben of Trove speaks about leading through that period.

> It was so frustrating because so many things were outside of our control. If you have a sense of the landscape, you can focus on how to be better within that context. If you can't figure out the rules of the game, you can't play well. So, then it just becomes complete luck. You can no longer rely on having good talent and making good decisions and being thoughtful and strategic—it's all out the window. It was really tumultuous.

The focus for leaders in true ambiguity is on lowering the uncertainty wherever possible. For the acute COVID disruption, the first one of these questions was "can our employees effectively work remotely?" For most companies, the answer to this question quickly became a resounding yes, and a new set of planning became possible. Another was "will a vaccine be developed and when?" As of this writing, companies have developed several highly effective vaccines, but global distribution is uneven.

Part of reducing that uncertainty is a ruthless focus on what your organization needs to do immediately to survive, particularly in the Respond phase of an acute disruption. Nick Scarpino of Portillo's notes this type of response at his restaurant chain: "For us, it's just been about let's control what we can actually control. In order to do that, we've had to strip away all the things that don't matter. It is incredible to me how fast our company has been able to pivot toward a curbside model, because we were able to just say okay 95% of things that we were doing were great ideas, but they don't matter now." As leaders gain clarity on these various questions and move to the Level 3 uncertainty, the scope of uncertainty narrows, enabling new types of planning for the future. As mentioned above, scenario planning can be an invaluable tool for helping address Level 3 uncertainty.

Scenario Planning

Scenario planning involves asking a series of questions about how the environment is likely to change in ways that affect your organization. Tom O'Toole, associate dean of executive education at Northwestern University's Kellogg School of Management and former chief marketing officer (CMO) of United Airlines and president of United's MileagePlus, emphasizes the importance of scenario planning: "The smartest companies, very early on and to this day, shifted into a mode of scenario planning with a best case, worst case, and base case. It enables companies to shift to a predictive but still flexible approach to the business as opposed to what I call speculation versus conjecture." The process of scenario planning involves asking a series of questions related to your business and possible future outcomes in relation to it:

- What is the new environment?
- What are the trends and implications?
- How might you regroup?
- How can you activate your plan?

One company we interviewed, Analog Devices, had conducted scenario planning several months before the acute disruption. The chief technology officer (CTO), Daniel Leibholz, describes the outcomes of that process.

> We actually went back to our scenario planning results about two months ago. We looked at what we landed upon in terms of the most important

mega trends, and things we needed to be preparing for. Honestly, a pandemic was not one of the things we anticipated, but its impact on the mega-trends has been one of acceleration. Some might be prioritized above others in the short term as a result of recent events, but none have been reversed.

What sort of trends did Analog Devices come up with in their scenario planning before the acute disruption? The digitalization of healthcare, increased factory automation, and adaptable and flexible supply chains were three trends that have gained momentum as a result of the COVID disruption. Leibholz reflects on two of these insights that emerged from their scenario planning. The first involved trends in healthcare. He explains that "one area that we thought was really important to digital health care was aging in place and managing wellness. This is accelerating. The trend of connecting high-quality sensors to your care provider continuously and autonomously is also accelerating, and it's coupled with a telemedicine revolution." The trends that the team identified in healthcare have simply been raised in importance due to the pandemic.

A similar trend occurred with respect to automation in manufacturing. He explains this trend, saying that "the challenges in the factory are not just around improving efficiency, but to add flexibility and resilience to supply chain disruptions." In fact, we'll identify exactly this type of application from Hitachi in the next chapter.

An outcome of the scenario planning process is that it forces leaders to intentionally and critically consider the changes posed by disruption. Even when the scenario planning understandably misses the biggest trend that will influence the coming year, the actions that the scenario planning process encourages remain viable. Even if one gets the scenario planning wrong or did not plan a certain scenario in advance, the process remains a vital component to dealing with uncertainty.

Kristin Darby of Envision Healthcare also speaks of the value of scenario planning:

> Even though we were dealing with a novel coronavirus, we relied on our collective expertise responding to other public health emergencies like Ebola. In this case, because the unknowns were much vaguer, the necessary level of readiness and contingency planning became much more critical so we had to

remain flexible as the crisis progressed. We also drew from our extensive experience caring for communities during other crises like hurricanes and natural
disasters to support local health systems in need of additional clinical expertise
like the use of virtual health.

Scenario planning is simply a mechanism for long-term strategic
planning, and the length one plans for determines how long of a timeframe to plan for. During the most uncertain times of true ambiguity,
timeframes may be measured in weeks or months. During the chronic
digital disruption, however, we encourage much longer-term planning
on a ten- to twenty-year timeframe. The same general methods and
principles typically apply in both cases.

Possible Strategic Moves from Scenario Planning

Once you have outlined the likeliest possible scenarios that your company
is facing with respect to disruption, how do you decide how to respond?
We recommend engaging in one of three types of strategic moves.

The first is to engage in *no-regrets moves*. As you engage in scenario
planning, it may be that one strategic action is applicable to multiple
possible foreseeable scenarios. Thus, you can engage in these actions with
relatively high confidence that they will pay off regardless of the current
uncertainty. For example, John Glaser, former senior VP of population
health at Cerner, describes these types of no-regrets moves for planning
in uncertain situations: "Work on things that are likely to be relevant to
many possible futures. Tell me a future in which engaging patients to
manage their own health is a bad idea, because I don't see that future at
all. So, I may not know how it's going to play out, but under almost any
conceivable circumstance, these things will be relevant."

Emma Lewis of Shell also emphasizes the importance of avoiding
decisions you'll regret. She says that "you've got competing short- and
long-term agendas. We've tried to look at where we want to go very
long term, identifying the short-term things we can do right now to
help ourselves that won't be a regret. Because you don't want to go fire
sale something and then find out three years later that was one of the
key pieces you need." Managers should push for no-regrets moves in

chronic disruptions as well as acute ones, because many of the same solutions stay the same with scenario planning.

The second is to engage in moves that create *options*. An option is a small investment that one makes now to create an opportunity to exercise a certain strategic action in the future. The concept is derived from financial options in which the buyer pays a small amount to preserve the right to buy a stock at a certain price level in the future. A real option is an inexpensive experiment conducted today that would enable us to take more action in the future. For example, Gensler had cultivated an option for remote work that paid off during the lockdown. Gensler's Joseph Joseph notes that

> because we already had been looking five to eight years in advance, we were able to flip the firm to work from home on the flip of a dime. We were probably among the top 1% in the industry that was ready. I attribute that in part to our passion all along for the importance of diversifying how we access our data and being prepared for crises that may be because of Mother Nature in certain geographical areas. That planning paid off with COVID.

The third move is to identify and engage in *big bets*. In the 2001 movie *Ocean's Eleven*, the character Danny Ocean, played by George Clooney, proclaims, "The house always wins. If you play long enough, and you never change the stakes, the house takes you. Unless, when that perfect hand comes along, you bet big, then you take the house." It may be that during scenario planning you identify a "perfect hand," a course of action that, should it play out, would lead to massive rewards. There may be a significant risk of failure in taking a big bet, but the payoff could represent a fundamentally different future for your company if successful.

Doug Mack, CEO and director of sporting goods apparel company Fanatics, describes the opportunity to make big bets.

> Ultimately, we expect a very favorable merger and acquisition environment coming in the future. Valuations will drop, and we'll be able to get some really interesting assets that we can add to our portfolio so that when we emerge from the pandemic, we will have even more capabilities than we've had in the past. So, we went out and did a virtual roadshow with Goldman we were targeting to raise $250 million, we ended up getting oversubscribed. We had to cap the round at $350 million. Within one week of closing in August, our first acquisition opportunity came along, a really wonderful company

within our industry had their bank come in and put them into liquidation
and create an asset purchase opportunity for us.

Fanatics was making big bets to take advantage of the disrupted environment to strengthen its portfolio of companies and capabilities.

Making Decisions in the Face of Uncertainty

Periods of acute disruption is a time when biased decision-making is more
likely to occur. The Nobel Prize–winning psychologist and economist
Daniel Kahneman notes that people typically exhibit two complementary,
but different, styles of thinking, which he calls System 1 and System 2:

- System 1 thinking is much faster and represents more routine types
  of cognitive tasks. These tasks are cognitively easier to manage because
  we have usually seen and dealt with them before. We develop shortcut
  rules, known as heuristics, which allow us to manage different scenarios and tasks more quickly.

- In contrast, System 2 thinking is slower and represents more deliber-
  ative types of tasks. Typically, these cognitive tasks are not performed
  as often and may require more conscious effort.[5]

The misapplication of System 1 thinking to System 2 problems results
in bias, and it is common during acute disruption when what would
have been a System 1 problem in normal times has shifted and become
different. Author and Babson College professor Thomas H. Davenport
offers a useful guide to the different types of bias that can lead us to
misapply the "tried and true" in the midst of disruption to situations
that require more creative leadership approaches.[6] Some forms of biases
he identifies are the following:

- Status quo bias: The mistaken belief that business will work the same
  way in the future as it did in the past because some aspects of the current situation seem familiar

- Confirmation bias: Finding information or evidence to support a decision one has already made or would like to make and disregarding
  information that goes against existing beliefs or decision preferences

- Availability bias: When people rely mainly on the first examples or information that comes to mind
- Bandwagon effect: When people opt for a particular decision simply because many others have made that particular decision

Many of the digital tools that have been proven indispensable for gathering the information essential to successfully navigating the disruption in the current technological environment can actually exacerbate these cognitive biases. For example, *confirmation bias* is particularly problematic in the age of the internet, where Google will readily provide you information to support your perspective, depending on which terms you enter into the search box. These tools also exacerbate *availability bias*, when people typically don't search beyond the first few results. Social networks such as Facebook and LinkedIn potentially exacerbate *bandwagon* effects by making users more aware of the decisions and actions of others. *Status quo bias* is inherent in any data-driven decision that may be based on past data. For example, one cause of the 2008 financial crisis was data-driven models that failed to recognize that the underlying risk conditions on which they were built had changed fundamentally. Therefore, it becomes even more important for managers to be critical thinkers and savvy consumers of data in times of acute disruption in a digital age.

We would add one more source of bias to Davenport's list when it comes to digital disruption—technical debt,[7] which occurs when significant investments made in a previous generation of technologies limit the strategic options in the current environment. It is a form of status quo bias, thinking that things need to be done a particular way because they have always been done that way. Processes are often modified or customized over and over as a result of new managers ("I want it done my way"), acquisition integration ("let's not mess with it"), or commercial constructs ("make it work for the sale").

Instead, managers should take the implementation of new digital tools as an opportunity to reevaluate their ways of working and opt for process simplicity alongside technology upgrades. Doing so not only ensures that companies are operating as simply as possible—which often is correlated with leading customer experiences—but positioning themselves to take advantage of technology updates and upgrades down the road. How digital technologies enable different ways of working is the focus of part II.

How to Apply the Concepts from This Chapter: Scenario Planning Simulation

Let's return to the exercise we started in chapter 1. You will now take the trends, uncertainties, and possible outcomes you identified there and begin to plan how to respond to them.

4. What are possible scenarios—or combinations of trends/uncertainties—that might come to fruition? Which scenario is likely, and which one to two scenarios are most extreme? (Note: The base case and extreme case will be the same for fairly predictable trends.) See question 4 for some example scenarios, trends, and uncertainties.

Question 4: Possible scenarios (sample)

Scenario	Trends	Uncertainties
Base case	• Rapid erosion of competitive advantage due to technology • Rise of ecosystems • Investment in technology • Increased consumer power	• Limited talent availability • Innovations in material science accelerate • Climate change is more front and center for companies • Economies return to a "new normal" as they adjust to the effects of the pandemic
Extreme case*		• Political tensions worsen • Economic power shifts east • Limited talent availability • Data ownership is the greatest competitive advantage • IP protections worsen • Markets become even more closed off as the pandemic continues to rage on
Alternate cases	• Rapid urbanization • Global proliferation of high-speed internet (e.g., 5G, satellite) • Rapid erosion of competitive advantage due to technology • Rise of the African consumer class	• Shift toward consumer privacy over a data-first mindset • More frequent and severe global health crises • Economic power shifts east • Continued attention toward social issues

*Trends for extreme cases are simply magnified versions of the base case trends.

5. What would need to be true to Respond, Regroup, and Thrive to these priority scenarios?

6. What no-regrets moves can we take today to best prepare? See questions 5 and 6 for examples of no-regrets actions for the Respond, Regroup, and Thrive stages.

Questions 5 and 6: No-regrets actions in different stages of disruption

Scenario	Respond	Regroup	Thrive
Base case	• Put proactive talent retention strategies in place • . . .	• Dedicate funding toward technologies that improve customer experience • . . .	• Develop and communicate a clear climate change goal and plan • . . .
Extreme case	• Identify parts of the supply chain vulnerable to political turmoil and develop contingency plans • . . .	• Identify, protect, and monetize data sources that could serve as a competitive advantage • . . .	• Deploy any technology that will drive consumer value and improve operating cost • . . .
Alternate case	• Action 1 • . . .	• Action 2 • . . .	• Action 3 • . . .

It is worth noting that scenario planning is inherently complex, ambiguous, and challenging. As such, we recommend a few practices based on organizations we have partnered with:

• Intentionally dedicate capital (talent, time, budget) to focus on emerging, unseen, or less understood trends and opportunities in the marketplace.

• Identify a few partners outside of your industry to explore innovation in science and technology. Consultants, software and cloud companies, and academia are good places to start.

• Commit the full management team to attend and participate in networking forums and learning sessions. The art of the possible ought to be a business-led exploration, but knowledge of technology is critical.

Apply learnings in real time to prevent "exploration" sessions from becoming management boondoggles. Some of the most growth-oriented companies test the art of the possible into action week after week.

II   How Digital Tools Help Organizations Operate in Disruption

5 Developing Your Digital Innovation Superpowers

We become what we behold. We shape our tools, and thereafter our tools shape us.

—Marshall McLuhan

The manager of a factory calls a repairperson to fix a complicated and expensive piece of machinery. The repairperson shows up on site, inspects the machine carefully, and restores the machinery back to functioning with one twist of his wrench. He then hands the factory manager an invoice totaling $5,000. The factory manager is outraged to be charged such an exorbitant sum for such a seemingly simple repair, demanding that the repairperson itemize the invoice to justify the expense. The repairperson then hands the factory manager another invoice that breaks down the expense as follows:

- Turning one bolt: $100
- Knowing which bolt to turn: $4,900[1]

What the repair person brought was not only the right set of tools but also the ability to quickly diagnose the situation and develop the appropriate solution. This parable highlights the difference between simply having a tool and being able to expertly wield that tool to solve a problem. In fact, having the right tool in and of itself might, in fact, get you into more trouble (as some of our spouses and partners can attest). Besides, why acquire a tool if you use it only once?

Technology Is Only Part of Digital Resilience

Many of us persist in the search for the right tools, particularly when it comes to responding to digital disruption. We tend to want to find the right "bright and shiny object" we can acquire and that, once acquired, will solve all of our problems and propel us down the path of digital transformation. We vehemently disagree with this perspective. Our perspective—that people are the real key to digital transformation—was shaped by more than five years of research involving thousands of organizations. It is a perspective that is often overlooked when companies seek to employ digital tools for new business purposes.

This desire to focus on finding the right tools is the equivalent to gravitational pull—an invisible, but often overwhelming, force. Part of this gravitational pull comes from an unintended consequence of top-down mandates, managers looking to satisfy CEOs or boards by exploring/adopting a single technology. Other times, managers simply want to purchase a simple solution to a complex problem. It's not that we think that the technological tools are unimportant but simply that they command a disproportionate amount of attention, oftentimes to the detriment of effective application and use of those technologies.

Having set out our perspective that technology is only part of digital resilience, it is also clear that certain technological investments have paid off better than others for companies seeking to adapt to the acute disruption. Therefore, we do think that it is useful in this book to spend time discussing the key families of technologies that help companies Respond to, Regroup from, and Thrive during disruption. Specifically, in the next three chapters, we look at cloud computing; data, analytics, and machine learning; and cybersecurity. In many cases, investments in the right technological infrastructure prior to a crisis proved to be the critical differentiators in determining how quickly and effectively companies responded when disruption struck.

However, it is not just about having the right tools. As the old adage states, "a fool with a tool is still a fool." But the right tools in the right hands can make the difference, and our objective here is to describe which types of technological investments pay off most richly in companies' responses to disruption. In fact, we think that the degree to which many companies were seemingly able to accelerate their digital transformation overnight powerfully reinforces our central hypothesis that organizational, managerial, strategic, and cultural issues are the barriers holding companies back from embracing a digital future, not technology.

The real question that managers should concern themselves with is what problems do the digital tools help your organization solve? What novel capabilities or strategic actions do they enable? How do those capabilities add value to your business model? Before we get into a discussion of the technologies themselves in the coming chapters, we want to focus first on these essential organizational capabilities that these tools enable for effectively dealing with disruption.

The Technologies We Pick Shape the Capabilities We Have

Based on our research and observations, digital technologies enable four key capabilities that we consistently find in digitally resilient organizations—nimbleness, scalability, stability, and optionality. While this list is neither exhaustive nor scientifically tested, the frequency of mention across our interviews suggests, at a minimum, that these capabilities be considered as a set of hypotheses that merits consideration in examining the anatomy of digitally resilient organizations.

These four sets of technology-enabled capabilities overlap and mutually reinforce one another. Nimbleness, scalability, stability, and optionality operate in concert. It may be useful to understand each capability using the following shorthand descriptions:

- Nimbleness: Rapid pivots ("we used to do this, and now we do that")
- Scalability: Rapid capacity shifts ("we used to serve x customers; we now serve 100x customers")
- Stability: Unlikely to give way under pressure ("our enterprise will persist despite continuing pressures and challenges")

• Optionality: Acquiring a net new capability by collaborating with another organization ("we identify underserved customers that we couldn't do previously due to our work with a new data provider")

Furthermore, these capabilities are relevant for companies adapting to different types of acute disruption. For example, Dorothy Leidner, Ferguson professor of information systems at Baylor University, investigated lessons for technology leaders following the dot-com bust and the 2001 World Trade Center attacks.[2] She and her coauthors identified several key lessons for technology leaders that generally parallel our concepts of scalability ("even out ups and downs"), stability ("ensure IT remains aligned with business goals"), and nimbleness ("yield business agility"), even though her article describes a very different technological infrastructure that is used to enable these capabilities than the ones available today. This suggests that although the digital technologies may change, the fundamental principles—nimbleness, scalability, stability, and optionality—remain constant.

Nimbleness

Technologies enable *nimbleness* on the part of digitally resilient organizations. Nimbleness refers to both the speed at which these organizations act and the ability to pivot when circumstances merit a significant change in direction. We have used nimbleness in lieu of agility since the latter term has lately been co-opted to describe a methodology that originated in the world of software development for organizing and executing projects and is now also used more broadly to describe non-technology project management approaches and techniques.

Nimble companies may use agile methods, but agile methods are not the only way to be nimble. Technology is not only an enabler of nimbleness with respect to acute disruption but can also be a forcing mechanism for nimbleness. As digital technologies enable the world to move faster, they also require that companies become nimbler to compete in it. Andy Ruben of Trove emphasizes the importance of nimbleness, saying that "in moments of incredible change, that's when the companies which are the nimblest thrive." The notion of rapid pivots is key to being nimble during acute disruption, such as remote work and contactless interactions.

Brad Surak, formerly of Hitachi Vantara, talks about how the digitalization of the company's factories allowed it to repurpose its safety and productivity systems into a social distance monitoring infrastructure, enabling its factories to get back to capacity without sacrificing employee safety. Hitachi has been able to repurpose its existing system of cameras and sensors that had originally been implemented to monitor safety and efficiency in factories to now also monitor both social distancing and possible COVID symptoms. Surak explains the following:

> Everybody has cameras and we've also been adding LIDAR and thermal imaging to it. Previously, these solutions have been driving productivity improvements, but we can then go in there and quickly add in a mask detection and social distancing algorithm. We ran a hackathon, and within two weeks now we've got the social distance analytics. We can automatically monitor how often you violated the two-meter restriction and provide feedback to employees.

When a robust digital infrastructure is in place and the organization has the right culture to leverage the capabilities it enables, it is relatively easy to repurpose that infrastructure for new challenges that companies may face during disruption. As we noted in chapter 2, the Chicago restaurant chain Portillo's was able to use its digital platform to shift its workforce from turning crew members into call center workers and delivery drivers. Nick Scarpino from the company explains that "we've launched an entire self-delivery program in the last few months. We're hiring about 150 drivers. We're using this as a strategic advantage, so instead of relying on a third party, you can have the true Portillo's experience the entire time."

Similar adaptation also occurred in higher-end restaurants. Alinea, one of Chicago's better-known restaurants with three Michelin stars, is frequently cited on lists of the top restaurants of the world. Alinea's website promises "We continue to push ourselves and our patrons to rethink what a restaurant can be." The restaurant lived up to this promise when it rapidly reimagined itself first as Alinea To Go (curbside

pickup from regional classics to six-course tasting menus) and then AIR
(Alinea in Residence: A rooftop dining experience in the West Loop).

Scalability

Another capability that digital technologies enable for digitally resilient organizations is *scalability*. The dictionary definition of scale is the
ability to change in size. Traditionally in business circles, "scalable" has
typically meant the ability to grow rapidly, referring to the growth of
companies such as Google, Facebook, and Uber. With respect to acute
disruption, however, scale does not simply refer to rapid growth in a
short time frame; it also refers to the ability to handle an unanticipated
increase (or decrease) of demand by many multiples overnight.

If a single technology brand is associated with the acute COVID disruption, it's Zoom. Usage of the video collaboration platform increased
by a factor of thirty between December and April 2020, going from ten
million daily participants to over three hundred million.[3] Of course, these
numbers are not unique to Zoom. Ian King reported in *Bloomberg* that
Cisco's "Webex daily meeting volume . . . more than doubled since the
beginning of March and expanded 2½ times from February. At peak hours,
volume . . . [was] up 24 times where it would be normally. . . . While
dropped connections and delays getting into meetings . . . increased . . .
Cisco's technology . . . just about kept up with the volume."[4]

Other types of organizations also needed to scale up quickly to meet
demand. Amazon, FedEx, and UPS are all examples of organizations that
have experienced huge demand spikes. Mark Onisk of Skillsoft noted
such a spike in demand for digital learning content. He reported that
"in many cases, [COVID-19] advanced the digitization of the learning
process by about two to three years with some companies. We've seen a
300% increase in the consumption of our product."

McDonald's is another example of a company using digital tools for
scalability. Chris Whitfield, who leads restaurant technology innovation
efforts at McDonald's, says that "curbside delivery is an example of a digital first channel that ultimately increases the capacity of our restaurants.

For the most part, during peak demand hours, we are supply constrained, not demand constrained, which is a good place to be. But it makes solving the incremental operational issues really, really valuable."

Many companies—such as those in the travel and hospitality industries—have been on the opposite side of demand swings. What has become evident during the pandemic is that even the best prepared companies may be challenged to respond instantaneously to dramatic demand changes, and how they communicate with customers, employees, and business partners can be as important as increasing or decreasing capacity. In these cases, the ability to scale down rapidly has been an important capability to weathering the downturn.

For example, Tom O'Toole of Northwestern University references the importance of data and analytics in the travel industry to help identify the level at which the industry should maintain services as well as areas where demand may be coming back to match capacity. O'Toole explains that "the first step was using an empirical approach to try to figure out how severely, and in what structured way companies needed to scale back their operations and their operating costs. Then the focus has really turned to discerning where demand may continue to exist and where demand is beginning to recover."

Another example is Hilton. Experiencing a 90 percent drop in demand, the company knew it simply could have laid off or furloughed employees. Instead, it developed agreements with companies that were experiencing a demand surge by which Hilton employees could get a type of preferred application status. Matt Schuyler of Hilton describes this effort.

> In the very early days of this crisis, we knew we were going to have to temporarily suspend operations across many hotels. We were also seeing the need for surge hiring demand in industries like retail among others. So, we created a pathway for our team members to look at opportunities at those companies, and essentially Hilton would vouch for them.

To accomplish this feat at scale, Hilton "reversed" their recruitment system, allowing Hilton employees to search for jobs at the partner companies at scale. Schuyler notes that the digital capabilities Hilton had developed before the acute disruption struck could be adapted to ramp up these new initiatives at scale in a relatively short time frame.

> We created a digital engine to allow our team members to find those jobs easily. Our recruiting funnel literally just reversed, and we pivoted our digital platforms to support it. The beauty of what we've built here over the years, made it really manageable to use the backdrop of our HR system, coupled with our partners in the recruitment and learning space, to essentially transform the operation to be outside in versus inside out.

At its peak, Hilton had over 1.2 million jobs on its systems that its employees could search for. Taking this action on behalf of their employees, the company could rapidly scale down its employee headcount while preserving its brand as a hospitality company as well as preserving its relationship with its employees with the hopes that they would be more likely to return when demand did.

Stability

Stability refers to the ability for companies to maintain operational excellence while nimbly pivoting and rapidly scaling. In its early days, Facebook encouraged its employees to "move fast and break things." In 2014, when it was rapidly approaching one billon users and still growing nearly 20 percent annually, stability became increasingly important for the company to move into its next phase of growth. In turn, Mark Zuckerberg changed Facebook's motto to "move fast with stable infrastructure." When talking about the change in an interview with *Wired*'s Steven Levy, Zuckerberg said with a smirk, "It may not be quite as catchy as move fast and break things, but it's how we operate now."[5]

You may remember back in high school or college chemistry when you were asked to do experiments with unstable chemicals that could react or decompose without careful handling. When we talk about stable organizations, we are talking about the metaphorical equivalent of not exploding or spontaneously combusting when exposed to shocks, including rapid demand fluctuations, as well as shocks such as climate catastrophes, economic instability, or labor unrest. With respect to acute disruption in 2020, stability involves the ability to use digital platforms to enable companies to respond to shocks by shifting to a more socially distanced environment while maintaining work productively.

Kristin Darby of Envision Healthcare notes the importance of stability in helping the medical group prepare for a surge of COVID-19 patients in early March 2020:

> We didn't know where the surges were going to occur, but we knew it was imperative to have a coordinated response. First, we focused on getting our back-office operations stable and making sure our non-clinical teammates could support the clinicians who were focused on patient care. By strengthening our back-office operations, we could support the clinicians on the front lines delivering care when and where it was needed most.

She points to an example where the IT team developed an app so that they could use tablet computers in intensive care units, with one camera to monitor the patient and the other camera to monitor the readings from the equipment. Referring to it as "electronic personal protective equipment," the app enables physicians and clinicians to monitor patients closely without needing to enter the room.

Stability can also refer to maintaining technological uptime while experiencing massive upticks in demand and usage. Noah Glass of Olo refers to this type of stability for his customers: "It felt like some days we were like Atlas holding up the restaurant world, and now we've made it through the first wave of this thing. I said to my customers that I want you to walk away from the end of this knowing that we are this reliable stable platform." Glass goes on to describe two short outages they had experienced—seven and eight minutes a piece during peak time. Although we are impressed that Olo was able to operate under such adverse conditions with such limited downtime, it was clear that those incidents bothered Glass and pushed his company to do better. He reflects that "the fact that we only had these two relatively minor outages was great because we were doing double the load that we ever had done before. Also, when we did get knocked down, we stood up to that and showed we're resilient."

Stability also derives from the ability to minimize the constraints of physical space. Chris Whitfield of McDonald's described an unexpected benefit of Automated Order Taking, an initiative that was only in its nascent stages when the pandemic struck. He described a challenge in many restaurants where the employee taking drive-thru orders typically stood in an area where beverages and desserts were prepared, creating problems in terms of physical distancing.

Stable organizations are the equivalent of the Timex watches of the twentieth century, the world's best-selling wristwatch in the late 1960s. Timex's advertising at the time featured one the country's most trusted newscasters, John Cameron Swayze, and the slogan "It takes a licking and keeps on ticking." Stable organizations also need to be able to take their "licks" while keeping operations "ticking" along as efficiently as possible.

Optionality

Optionality is the ability to integrate another organization's capabilities to become even more nimble, scalable, and stable. The term is rooted in the concept of financial options that we referenced in chapter 4—spending a small amount today to provide the option to take a strategic action in the future. Optionality is often created through a robust ecosystem of partnerships that allow companies to leverage the capabilities of other organizations when needed. These partnerships are essential for helping companies adapt to both acute and chronic disruption.

The term *ecosystem* is rooted in ecology and the work of early twentieth-century botanist Arthur Tansley. It originally denoted a community of biological, chemical, and physical components that function as a unit; a beaver pond is an oft-used example. Youngjin Yoo, professor of entrepreneurship and information systems at Case Western Reserve University, describes a metaphor: "You could say that the company at the center of an ecosystem builds the beaver dam, and that, in turn, creates a pond that attracts other creatures, who also thrive there. The key strategic question is which strategic layer will the platform company control and which ones will it open to others?" In the business context, *ecosystem* has come to mean a group of companies that collaborate to achieve shared goals, with or without formal ties. While a digital ecosystem can include traditional partnerships and consortia, the term covers a wide array of relationships with external organizations and people. These include academic institutions, government entities, nonprofits, start-ups, customers, and even competitors.

It is important to have those relationships established in advance so that they can be leveraged when needed. The chief digital officer of a large, global insurer notes that need: "You really need to have already figured out in advance who are the partners you want to work with. I wouldn't want to be starting from scratch, trying to figure out who does the sort of thing we need." Colin Schiller, CEO of the virtual mentoring site Everwise, described this process: "We've had a bunch of companies come back to us, saying 'we talked to you in the past, and we felt then like we needed people live and on site. Given the way things have changed, however, virtual is actually very appealing now.' The contracts we lost as a result of the pandemic have been more than replaced by these new customers."

Nearly 80 percent of respondents to our earlier survey on digital disruption believe partnerships with other organizations are vital to their innovation efforts. Of course, as any frustrated dieter can tell you, believing something and acting on that belief are two different things. Companies at early stages of digital maturity are particularly less willing to commit resources to innovation. At the more advanced digitally maturing companies, 80 percent of respondents say their firm is cultivating innovation via partnerships. At developing firms, that number drops to 59 percent, and at early-stage companies, it falls further still to 33 percent. Digitally maturing companies are (thus) more than twice as likely to work with external organizations to innovate compared to the least digitally mature businesses (see figure 5.1).

These digitally maturing organizations use partnerships to support multiple dimensions of the innovation process and emphasize wide-ranging, capability-building ecosystems that address both short-term and long-term objectives. Part of the reason for this emphasis, says Yoo, is that ecosystems enable organizations to operate more flexibly, providing access to more collaborators and potential innovations. Frank

Leaders across all maturity levels recognize the importance of partnerships to innovation, but only at the maturing level are organizations consistently cultivating them. *(% of respondents who agree or strongly agree)*

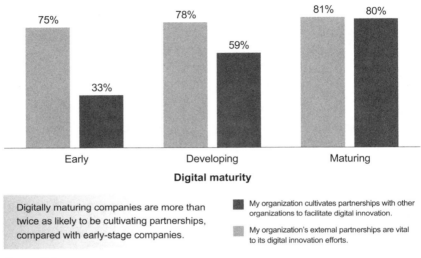

Digitally maturing companies are more than twice as likely to be cultivating partnerships, compared with early-stage companies.

■ My organization cultivates partnerships with other organizations to facilitate digital innovation.

▨ My organization's external partnerships are vital to its digital innovation efforts.

Figure 5.1
Partnerships and digital maturity

Nazzaro of Freddie Mac underscores this point: "It's just a healthy way to have partners on the street that are exploring new ideas. It's also a cost-effective way to innovate because they bear the burden of experimentation and then we can capitalize on it when it's at certain stages or we have certain needs."

Our interviews with corporate executives reveal several ways that these types of ecosystems feed innovation. Brice Challamel of Google Cloud indicates that forming these partnerships is an important part of digital transformation: "It's a journey with different phases. There's a first phase in which we need alignment on a vision for the future. The second element is the value model that we're creating together. And the third element is the change management roadmap for the ecosystem, which supports the need to evolve in order to deliver those solutions and unlock that value model."

Rajeev Ronanki of Anthem envisions an ecosystem of providers unified by a shared data infrastructure that may emerge from acute disruption.

> We know that care delivery can be better with a combination of digital, virtual, and physical. I think we can also create a more connected provider ecosystem where each provider system in our network is peripherally aware of every other node in that network through data. And so, in so doing, you can basically envelope the country with a set of digital and AI solutions that would serve as well with the next crisis, but also just improve the way care gets done, period.

Ecosystems also contribute to innovation through their collective access to diverse customers. In online video software development and distribution, Dave Otten, CEO of JW Player, says that technologies like JW Player come together with advertising technology companies and advertising partnerships. All of them, along with the audiences they gather from across the broader web, form a collective ecosystem. Their collective understanding of audience feedback and behavior can play a critical role in the innovation process.

Working with external partners, however, also presents difficulties. We asked survey respondents to share their biggest challenges with leveraging partnerships and networks to increase innovation. Nearly half (46 percent) of all respondents cite challenges related to creating a collaborative culture and to aligning goals across an ecosystem. When it comes to culture, companies struggle with employees and leaders who aren't naturally inclined to collaborate with external partners.

This problem deepens when trying to develop goals that are acceptable to all parties within the network. JW Player's Otten acknowledges the challenge but advises companies to "balance the need to hold onto the core of your culture, while letting go of the things you need to in order to grow up." When companies are navigating these issues, Amy Smith, VP for corporate engagement at Techstars, notes that "executive leadership is really, really important" to address these issues. Leaders ultimately create credibility around enterprise strategy that facilitate participation in the ecosystem.

How to Apply the Concepts from This Chapter: Creating Optionality through Ecosystems Framework

Although digital technologies can facilitate optionality, companies often struggle with articulating the ecosystems they wish to draw value from. The fear of the unknown is most often what prevents established organizations from diving in. Ironically, these companies often have the most to gain from engaging in a robust ecosystem. We have had success helping executives and organizations define the ecosystems they wish to play in, their role in the ecosystem, and how to create and extract value from the ecosystem by guiding them through four questions. You can use these questions on your own to consider how to develop robust ecosystems and place your answers into the ecosystem map in table 5.1.

1. What is the **objective** of the ecosystem we wish to create/participate in?

The best objectives are ones that are broad, solve a human need, and require multiple players to participate. For example, when building a mobility ecosystem, the objective can be as simple as "move people, goods, and services between two points." Naturally, no single company can do this alone, creating the need for an ecosystem of players that bring new and differentiated value to the it. Another example is to "improve people's physical and mental well-being."

2. What **needs** are we solving for? Which ones are met, and where are there **gaps**?

The next step is breaking down the objective into a set of customer/ user needs that articulate the outcomes the ecosystem will deliver. Here, it is critical to think like a customer (versus a seller or ecosystem partici- pant). Voice of the customer studies, talking to the frontline sales force, or engaging outside help are particularly useful here. In the mobility example, the customer needs could be their daily commute, long- distance transport of goods, business travel, or similar. For well-being, the needs might be around diagnostics, preventative care, emergency response, etc. Use the columns of the table to complete this question.

Table 5.1
Mobility ecosystem map

(1) Ecosystem objective: Move people, goods, and services between two points		(2) Needs				
		(2a) Daily commute	(2b) Business travel	(2c) Long-distance goods transport	(2d) Last-mile goods transport	(2e) . . .
(3) Capabilities	(a) Vehicles	Deliver	Deliver	New partner—Daimler	Deliver	
	(b) In-transit experience	Existing partner—Pandora, Sirius XM, Spotify New partner—TBD		NA		
	(c) Fleet mgt	New partner—Lyft/Uber		New partner—FedEx/UPS		
	(d) Energy	New partner—TBD		New partner—Shell		
	(e) Payments	New partner—Mastercard New partner—Stripe				
	f) . . .					

3. What **capabilities** are required to address unmet needs?

Like we introduced in chapter 2, capabilities describe the set of objectives, processes, technologies, and talent that collectively generate value for an organization and allow it to deliver its strategy—this time the consumer needs. The idea here is to think at a high level and not get bogged down in the details—there will be time for detail later. For example, to deliver against the mobility objective, an ecosystem will need participants who make cars, others that can support fleet operations, and still others that can provide a differentiated in-vehicle experience. The ecosystem could go so far as to include participants who provide digital infrastructure (e.g., connectivity), energy infrastructure (e.g., electric charging, fuel), and enablers (e.g., mapping software). Use the rows of the table to complete this question.

4. What capabilities can we **deliver**? Whom can we **partner** with to deliver value?

The mobility ecosystem map in table 5.1 creates a canvas from which to choose where companies can contribute. In most instances, enterprises will have a few "intersections" covered. For example, if General Motors (GM) wished to participate in a mobility ecosystem, it already has vehicle capabilities as well as existing partnerships with companies that offer in-transit experiences. As the ecosystem gets further built out, the need for new capabilities will emerge, which are further away from the players' core capabilities, requiring strategic partnerships—bringing the ecosystem to life. Indeed, the ecosystem can include a wide-ranging set of players from payment processors (e.g., Mastercard) to connectivity providers (e.g., Verizon) to electric vehicle charging—the possibilities are endless. As such, it is important to have a way to evaluate new opportunities and to allocate dollars and resources only toward the most promising ones. When choosing partners and inviting organizations to the ecosystem, remember it is critical to have a set of principles that all participants can agree on. These include the objective of the ecosystem but also how new partners are invited, how value is shared, and how the ecosystem will evolve. The world is littered with examples of novel ecosystems that failed because of a lack of governance—without structure, participants may overreach, lose interest, or, worse, jeopardize your customers' experiences.

There are a few pits that companies often fall into that often prevent ecosystems from realizing their full potential. The following are the most common to keep in mind when conducting this exercise:

- **Defining the ecosystem too narrowly,** limiting the ability to engage a variety of participants, ideas, and solutions
- **Defining the ecosystem too selfishly,** solving for their own problems instead of the customer's problems, or ones that an ecosystem can truly support
- **Trying to control the ecosystem too tightly,** limiting the flow of information and value between players, which disincentivizes participation or creates silos
- **Turning the ecosystem into a value chain,** with defined suppliers and distributors, instead of allowing participants to play multiple unique roles

As you complete this exercise, you can start to think outside the organization's boundaries and consider how these partnerships can not only create optionality but also enable nimbleness, scalability, and stability through a robust ecosystem of partnerships.

6 How to Move at Cloud Speed

There was a time when every household, town, farm or village had its own water well. Today, shared public utilities give us access to clean water by simply turning on the tap; cloud computing works in a similar fashion. Just like water from the tap in your kitchen, cloud computing services can be turned on or off quickly as needed. Like at the water company, there is a team of dedicated professionals making sure the service provided is safe, secure and available on a 24/7 basis. When the tap isn't on, not only are you saving water, but you aren't paying for resources you don't currently need.

—Vivek Kundra

In 2003, *Harvard Business Review* editor Nicholas Carr penned a bombshell article entitled "IT Doesn't Matter."[1] His argument was that the hard-core tech of the IT revolution would be relegated to the role of a ubiquitous commodity and no longer be a source of competitive advantage for companies. Strongly held arguments exploded from both sides of the divide.

Nearly twenty years later, we realize that Carr was both right and wrong. He was right in that simply owning more technology is not, in fact, a source of competitive advantage. As Carr predicted, computing has become ubiquitous and can be rented as needed with very low out-of-pocket costs. This trend is often referred to as cloud computing, and it is one of the hottest frontiers in technology today. While writing this book, Snowflake, a cloud-based data-warehousing company, launched the biggest initial public offering of all time, with an opening valuation of $70 billion. Some might even argue today that trying to own and manage all of your own technological infrastructure, regardless of where and how that technology is deployed, actually puts a company at a substantial competitive disadvantage.

Carr was wrong, however, in concluding that IT somehow does not matter as a result. The number of different capabilities and services that are now available through cloud computing platforms is diverse, customizable, and combinable; together, these various cloud components make IT matter more than ever. Today, technology applied well to business problems and opportunities ought to be a substantial competitive advantage.

Although the examples in this chapter focus primarily on how cloud computing helps companies respond to the acute COVID disruption, it also provides virtual mission continuity for all types of acute disruptions by ensuring an organization's digital capabilities remain operational and accessible, even if other aspects of the organizational or societal infrastructure are disrupted.[2] Colin Schiller of Everwise described the benefits of cloud computing: "When we started this business, everything had already gone to the cloud. So, it was actually shocking to me from a technology perspective how little we had to change to seamlessly shift. Other than a few people needing to upgrade their home Internet connection, it was seamless . . . shockingly easy."

Cloud Computing Is a Matter of Degree

In 2011, Andrew McAfee called cloud computing a "sea change—a deep and permanent shift in how computing power is generated and consumed."[3] At the risk of oversimplification, cloud computing involves renting computing power, storage space, and other digital resources from another company rather than running it on your own computers, servers, and network. Although many speak about cloud computing as if it is a single thing, there are actually three different levels of cloud computing. The specific terminology can get confusing, so it may be helpful to think of the different types as a house:

1. The most basic type of cloud computing is called infrastructure-as-a-service (IaaS), which consists of the rented data centers and computing power for basic technical computing resources like servers, storage, networking, and databases. In our house metaphor, this would equate to renting an empty house, and you are responsible for furnishing the house yourself. For our purposes, we include in this category another type of cloud computing called platform-as-a-service (PaaS). The focus for both IaaS and PaaS is on acquiring technological capabilities. The distinction between the two is not important for our purposes, so we will just combine them for simplicity. There is considerable overlap between leading companies for both IaaS and PaaS—Amazon Web Services (AWS), Microsoft Azure, and Google Cloud.

2. The second level of cloud computing is known as software-as-a-service (SaaS). These are mostly ready-to-use cloud-based applications that you can mix and match, depending on your needs. SaaS applications have varying degrees of enterprise readiness when implemented and adopted, allowing for some configuration, though typically not full customization. This is the equivalent of a fully furnished, stocked, and staged house. For SaaS, you are acquiring both technology and business process capabilities, and many SaaS companies actually rely on IaaS providers. Leading companies in this space are Salesforce, ServiceNow, Slack, and Workday. Almost all enterprise software providers now offer SaaS versions of their products.

3. The final level of cloud computing is business process-as-a-service (BPaaS). These are cloud-powered business tools like OpenTable for restaurant reservation and capacity management that allow companies to package custom services and consulting in addition to the software offering. In this part of the metaphor, you are only renting the specific rooms that you need, and it comes with concierge service. Here, you are acquiring technology, business processes, and the people to support them.

Now that we understand the basic technological aspects of cloud computing, we can move to a discussion of how the cloud supports the four characteristics of stability, nimbleness, scalability, and optionality. We elevate stability to the top of this list because we think the stability that the cloud provides is the most critical benefit with respect to acute disruption.

Enabling Stability through Best-in-Class Expertise

Ironically, stability was initially one of the major concerns with cloud computing. Although SaaS companies like Salesforce already existed, the potential of cloud computing exploded in 2006 with the launch of AWS, where it was met with significant resistance among legacy companies. Would it be reliable and secure? Can I trust my company's valuable data to someone else? What happens if the provider experiences an outage?

Many of the initial questions about the reliability and security of cloud have worked out in favor of the cloud as it has matured. Eric Ranta of Google Cloud emphasizes this evolution in the acceptance of cloud computing: "Cloud is now pervasive enough where people have talked about it for a while, so there's not a fear factor that comes into play. When I started at Workday in 2012, you had to convince people why they would want to keep their data outside of their four walls. We don't have those same discussions anymore."

Today, reliability is one of the key features of cloud computing. The goal for cloud computing companies—particularly IaaS and PaaS providers—is to provide 99.999 percent reliability ("five nines"), and most come remarkably close to this target. What may be more impressive about the cloud is that given the capacity available, you may never need to actually take an application down. The capacity itself at least will never be the constraint to application availability.

Interestingly, stability doesn't necessarily arise from anything inherent in the technology behind cloud computing, but it arises in the ability to access the expertise of those who manage that technology. At the foundation of many cloud services are the leading global technology companies—Amazon, Google, and Microsoft. In a situation of acute disruption where unprecedented flexibility and demand shifts required best-in-class technology, cloud computing enabled companies of all sizes and resources to access those world-class resources.

The reliability numbers do not always translate to SaaS and BPaaS companies, but many strive to offer similar levels. For example, in chapter 5, we mention Olo's seven to eight minutes of downtime on two separate occasions during Friday night rushes at the peak of the acute disruption. The failure did not come from the AWS platform that hosted Olo but instead came from the complexities arising from certain combinations and permutations of menu customization options during peak load time, a problem the company has since identified and fixed. Noah Glass describes the problem: "The thing that got us was not the number of orders, we were prepared for the volume. It was the permutations of baskets and menu item complexity. That complexity, magnified by the order volume and number of baskets that were being created, took our system and us by surprise, but it could have been a lot worse."

In addition to providing security, the cloud also delivers enhanced security. Large technology companies are more likely to be aware regarding the latest security risks and act quickly to mitigate those risks and have security advantages that few small companies can match. For example, since Google monitors much of the world's web traffic, it can sense when hacking attacks are being attempted, giving it a lead on addressing those security concerns.

Tom O'Toole of Northwestern University emphasizes the security benefits of the cloud: "Skeptics might say we can't possibly move our systems to the cloud, because cloud isn't secure. I would push back on that. I would probably put the data security of AWS or Microsoft Azure against virtually any internal enterprise system I've seen." Regardless, the shift to the cloud does not necessarily change a company's approach to cybersecurity, because often users are the weak links, not the technology. We will address cybersecurity in greater depth in chapter 8.

Another fringe benefit of stability provided by cloud computing is more environmentally sustainable. Although data centers use massive amounts of energy, they have significant financial incentives to lower energy use. For example, Google was able to lower the energy usage in one of its data centers with the implementation of AI to manage usage. This relentless focus on efficiency is at the heart of the cloud providers; AWS and Google also have their equipment custom built based on specs to meet their performance requirements. They can make strategic decisions on where to build the data centers, such as in areas where sustainable power is more likely (e.g., near hydroelectric dams). And they can run more efficiently, only powering what is needed at the time. Electricity is a major cost for these data centers that power the cloud, and firms have a strong economic incentive to make them as efficient as possible.[4] The amount of computer processing grew by 600 percent between 2010 and 2018, but the overall usage of electricity grew by only 6 percent.

Enabling Optionality by Quickly Adding New Applications

Cloud computing also enables organizational nimbleness through the ability to rapidly adopt and implement new technology. Because it is relatively easy to begin using technology that is hosted in the cloud, it is much easier to begin exploring and using a technology when needed. Shamim Mohammad of CarMax underscores the importance of nimbleness enabled by cloud computing: "We have had a cloud-first mindset for several years. We're not locked into one technology, and we can add different types of cloud services into our platform. That level of flexibility, really, really helped us to be able to move quickly during this pandemic."

In many ways, one might think of cloud computing like a big Lego set of computing. Well-designed clouds have a myriad of technical Lego blocks (all shapes, sizes, and colors) that you can combine into pretty much anything you could want or need. Companies can mix and match functionalities, allowing them to participate in a broader ecosystem of innovation and services. This environment allows collaboration to become more borderless, because many cloud platforms work to improve the compatibility and interoperability among services on its platform. This ability to quickly add and adapt the technological infrastructure is the key feature that allows companies to be nimble in the face of disruption.

While cloud can be orders of magnitude easier than managing one's own IT infrastructure, it does take considerable effort and experience to obtain the level of expertise and mastery to achieve something that resembles plug and play capability; this is essentially the holy grail that many large enterprises struggle with as they navigate their technical debt burden. Nevertheless, once achieved, a robust cloud capability allows companies to rapidly conserve resources, reform strategy, have new market opportunities, and diagnose the permanence of disruption. Some moves may be temporary, while others may be long term or permanent.

For example, restaurants were forced to switch their business model literally overnight, pivoting from in-person dine-in to takeout and delivery. SaaS platforms like Olo allow restaurants to sign up relatively quickly, requiring only minor tweaking of the service to customize it for the particular business, such as the various menu items and combinations and restaurant locations. The platforms allow restaurants to rapidly shift their business model to focus on delivery, because a reliable technological infrastructure can be quickly deployed via the cloud to support this shift.

Cloud computing also allows Olo itself to quickly adapt its business model through its IaaS provider, AWS. Before the acute disruption struck, Olo had employed a dual strategy—supporting in-person dining as well as remote. It was able to shift its resources thanks to its cloud design and to support the rapid scaling of one part of the business and the ramping down of another. The ability for Olo to rapidly change and reprioritize its service offerings allows it to better support its customers in what represents—in many cases—an existential threat to their business. Additionally, cloud-based systems allow more rapid integration with partners because much of the work has already been performed and simply must be tailored to the existing environment. For example, Nick Scarpino of Portillo's notes that they were able to add two new delivery services—Uber Eats and Postmates—within a few weeks to better handle the increased demand for delivery.

Providing Scalability through Rapid Expansion and Contraction

During acute disruptions, companies can use the cloud to rapidly adopt or deploy new technology to serve remote working needs, quickly build new environments to capture new digital channel market opportunities, scale up existing technology to improve the capture of digital channel market opportunities, and scale down or turn off dormant or nonessential technology for business degradation. In other words, the cloud provides the agility necessary to rapidly pivot away from suspended business activities to areas that have greater consumer demand. Joseph Joseph of Gensler notes this benefit of cloud computing: "We already had an existing virtual infrastructure. Frankly, I say this in the humblest of ways, we'd already been very cutting edge in our strategy. So, we already had the infrastructure somewhat set up. All we have to do is turn all of this on, and we were able to transition the firm to remote work on the flip of a dime."

This ability to shift to remote work more quickly also enables a more flexible workforce. When employees do not conduct work in a physical space, it creates an opportunity for whom they can involve in the company's work—from customers to partners to outside experts. It also creates opportunities to adapt, change skills, eliminate certain work when not needed, and bring it back quickly when it is, particularly technical work. Many of the cloud technologies are enabling the remote workforce to thrive in a disruptive world, and the cloud itself creates conditions where businesses can not only be nimbler technically but also nimbler with their technology and business operations. While researching this book we have been able to work with a wider variety of companies to help them navigate the disruption, simply because the collaboration platforms provide a greater variety of ways to engage quickly and easily.

Many companies made the radical and immediate transition to remote work relatively seamlessly, and this seamlessness is due in no small part to cloud computing. Eric Ranta of Google Cloud notes that

> it's about being able to access your data or applications and all the things that you would do in a normal office environment remotely, in a secure performance, and quite frankly elegant manner. The lockdown exposed

companies that have a data center that is not fully equipped. Companies oftentimes underestimate the difficulty here. How do you make sure that you've got applications that your employees can use in any way, shape, or form? How do you make sure that you can do it in a secure and compliant way? How do you do that when somebody is competing with bandwidth in their own house with three other people? Some of these get down to very, very practical aspects.

Another example can be found in the airline sector. With reservations down nearly 90 percent, it required a major, immediate scaling back of loyalty, booking, reservations, and asset management systems; without the cloud, many of those airline companies would be stranded with infrastructure depreciation that they could have potentially scaled down. Reductions in flight volumes over the course of the next decade may require reductions in infrastructure and data capacity necessary for things such as maintenance logs and flight logs, which is essential for survival and self-preservation. The ability to pivot and redirect is also an important aspect. If on premises sales or cash payments are no longer important, companies can redirect resources away from the technologies underlying those aspects of the business and redirect them into areas of growth and sustenance.

Enabling Nimbleness through a Shift from Fixed to Variable Costs

Some of the nimbleness enabled by cloud computing also stems from the optionality it creates—quickly adding new capabilities and partners enhances a company's ability to move quickly. It also provides nimbleness by shifting the cost of IT from fixed to variable. In cloud computing's early days, many wondered whether it would cost more relative to companies owning their own technology. One of the financial implications is the shift of the cost structure for companies from potentially massive amounts of up-front fixed costs needed to set up a proprietary IT infrastructure to almost entirely variable costs, where—with balanced pricing and contract strategies—you only pay for the amount of technology you actually need and use. If done right, this can result in less waste and technology leakage, as most companies suffer from unused and underused technology. This ability to pay for only the computing power you use enables optionality, the ability to add or remove capabilities as needed.

There is a significant benefit for relatively small companies because they can often have access to world-class technology without huge up-front fixed costs of building a proprietary IT infrastructure. As companies grow, however, the costs tend to balance out. Of course, what you get for that cost is usually a far superior technological infrastructure than most companies could possibly develop themselves.

With proper management and governance of the cloud environment, this shift of IT becoming a variable cost virtually eliminates the benefit of scale that was essential in the previous generation of IT.[5] Large companies used to have the advantage because they could maintain a large computing infrastructure. One of the drivers of consolidation in the healthcare industry over the past decade has been associated with the costs necessary to manage HIPAA-compliant information infrastructures. The IT infrastructure is so expensive and labor intensive to manage that only large companies can effectively manage it. Cloud computing renders that infrastructure cheap and available to all, making scale no longer an advantage but potentially a liability. Firms that manage their own data centers aren't simply competing against other companies in their industry; they're competing against the big tech players—Alibaba, Amazon, Google, and Microsoft. Firms now need to be lighter and agile, and the cloud both causes and provides that opportunity.

Furthermore, the cloud reduces the need for fixed costs since companies don't need to maintain as much IT staff to manage it. For example, Portillo's, the Chicago-based restaurant chain, has zero dedicated programmers, and yet it was able to execute some extremely agile changes in response to acute disruption thanks to its cloud providers. The financial services firm Robinhood has a similarly thin IT staff. These observations are only possible with heavy outsourcing and cloud use.

Are we advocating that companies get rid of their technology staff? Absolutely not. We advocate that companies rethink the operating model between business, technologists, and the cloud as the new IT partner. They can bring business and product leaders closer to internal technology and enable third-party cloud players to increase collaboration. The goal is to establish a core IT function that provides consistent and efficient shared services to the business. Cloud computing allows companies to rely considerably on digital technologies without needing to have the technologists on hand. When you move to the cloud, you typically get access to the world-class engineers who are working at the largest technology companies in the world.

Small and Large Companies Benefit Differently from the Cloud

For small companies, the cloud helps them start up with a robust technology infrastructure much faster, allowing them to punch above their weight.[6] Smaller companies greatly benefit from SaaS and BPaaS, where they can scale "business-ready" apps overnight to compete with larger companies. Many local food brands have achieved scale by using AWS for customer engagement and marketing in a way that large consumer packaged goods companies cannot or have not. Several companies that started small on AWS—such as Dropbox and Lyft—ended up getting large after launching and growing their businesses on cloud.

This ability for the cloud to enable smaller companies to operate stronger earlier creates several competitive challenges for other companies. Strategically, this means large multinationals can face competitive threats from a massive number of very small businesses that can ramp up competition quickly. For example, large numbers of fintech start-ups are seeking to disrupt every facet of inefficiencies in the financial services sector. The sector attracted nearly $35 billion in funding in 2019, providing nuanced approaches to banking, payments, insurance, currency, and lending.

A shift in venture capital funding as a result of cloud computing is accelerating this challenge.[7] When cloud computing makes it significantly less expensive to start a company, venture capitalists (VCs) can invest in a wider variety of companies—an approach colloquially known as "spray and pray." VCs end up funding lots of small experiments, resulting in an innovation infrastructure seeking to test every aspect of the business model. We expect this competitive threat to be exacerbated during times of disruption. As entrepreneurs start up new ventures in the wake of disruption, the competitive landscape for legacy companies might look quite different in the coming years.

For very large companies, cloud computing allows them to achieve efficiencies and effectiveness at scale very quickly. It increases the organization's nimbleness, speed to market, and access to new innovations that cloud computing enables, allowing companies to grow more effectively than business as usual.

Speed is a very important aspect of this equation. The ability to respond quickly to favorable economies at a micro or macro level by implementing services in a particular geography is key. Some projects are so massive that even large companies cannot do without large, distributed computing networks, such as autonomous vehicles, big data, and Internet of Things (IoT). Another strength of the cloud is the ability to quickly get small and redirect resources toward growing markets vis-a-vis technology. Peak-to-trough cycles are historically very painful financially, and the cloud can help bridge the downside of those cycles and also extend the upside by capturing more market share and enabling real growth.

How Does Cloud Computing Help Companies Manage Disruption?

Returning to our Respond, Regroup, Thrive framework introduced in chapter 2, cloud computing helps at all stages of the digital resilience process. It helps companies move through the Respond phase faster, because the technical infrastructure is in place to make the shift happen, providing capabilities to weather the initial shock of disruption more easily. It provides optionality necessary to add new capabilities to enable a pivot, facilitates scalability to enact that pivot at scale and to handle sizable demand shifts, provides stability by diverting the necessary resources to bear securely, and enables nimbleness by only paying for the services companies use.

Cloud computing also facilitates the Regroup phase. As companies are regrouping and identifying opportunities created by the disruption, a robust cloud infrastructure gives leaders the confidence that the technological capabilities will be in place to capitalize on those opportunities. Cloud computing also helps enable a growth mindset, because the tools to help companies develop organizational competencies to adapt are readily available to help. When other companies are moving to the cloud, more competitive pressure will be created because it also lets competitors be more fluid and adaptable.

Last, cloud computing can help companies in the Thrive phase by allowing them to scale their successes more rapidly. Through the experimentation that the cloud facilitates in the Regroup phase, companies can capitalize on successes rapidly. Many companies spend this early phase discovering which aspects of the business are less valuable in the disrupted world and which may be more valuable. When they discover which is which, the cloud lets them execute more quickly and capitalize on that knowledge. In other words, cloud computing allows them to shift from the Regroup to the Thrive phase far more quickly than they would be able to otherwise, letting them execute only when the way to thrive is clear.

The Golden Age of the Cloud Is Yet to Come

We think that the acute COVID disruption might actually propel cloud computing into a phase in which it thrives. Despite the myriad of benefits provided by cloud computing, Eric Ranta of Google Cloud believes that its most significant promise still lies in the future.

> I would say the golden age may still be ahead of us. Many companies now can answer the why of cloud computing, that part of the problem is very clear, but the how is still daunting to them. How am I going to move these mission critical workloads and applications and data into a cloud environment and then be able to get even better things out of it, meeting all the regulations, and be more secure and compliant than we've ever been before?

When companies can get past the daunting *how* of transitioning to cloud computing, it begins to open up the *what*—the new opportunities that the cloud enables them to do business differently. Ranta continues:

> When you ask senior executives where they want to be 5–10 years from now, they define themselves by some sort of aspirational type company or user experience that's much greater than what they are today. That will be the golden age of the cloud. We can take that strategic enablement, we can show them how to get new revenue, new business models, security, compliance—at a superior cost model. That will be the golden age of cloud. We may not be far from it but we're not there yet.

How to Apply the Concepts from This Chapter: Questions to Guide Cloud Adoption

Before starting, please dust off your activity from chapter 2, where you identified several capabilities important to your organization and classified them as strategic, core, and foundational.

How Do I Know Which As-a-Service Model to Choose?

There is no silver bullet when it comes to choosing the right as-a-service model; however, there are some tips and tricks that you can apply to ensure that you are making the right decision for your organization. At its core, choice is all about balancing the risk/reward trade-offs with strategic importance. Chapter 6 provided some clear examples regarding how the cloud is significantly affecting end users, customers, and business models, but we would like to bring that back to the heart of what an organization does—via delivery of its core capabilities.

When navigating cloud choices, some general guidelines can be applied to strike the right balance—you may have reached these conclusions by going through the thought exercise above with the three selected capabilities:

- **Foundational** capabilities are excellent candidates for BPaaS or SaaS solutions. These solutions enable considerable stability by relying on best-in-class processes and technologies. They may, however, reduce nimbleness because they are typically undifferentiated and offer limited customization since a third party has created software and often a service to fulfill the capability requirement.

- **Core** capabilities are candidates for SaaS but oftentimes with higher levels of customization to meet the core requirements. With core capabilities, the primary value is in the optionality it provides by allowing a certain degree of customization. Because of the importance of core, organizations don't often acquire the full-service maturity capability via BPaaS.

- **Strategic** capabilities are good candidates for IaaS. Nimbleness is the primary value of the cloud for strategic capabilities. When you don't need to worry about maintaining a data center, you can focus your energy on customizing and adapting solutions to changing market conditions. If clients choose to buy off-the-shelf SaaS or BPaaS solutions for strategic capabilities, they often heavily modify, customize, or configure the solution to meet their unique business requirements to the point that they may not be recognizable. They may also put significant intellectual property protections in place with their provider to ensure they are not leaking any of the secret sauce that makes them unique.

Update your capabilities matrix from chapter 2 with a current mapping of which technologies are primarily serving the capability. Are any of these capabilities being managed by in-house technology that could be migrated to a cloud environment? Investigate leading cloud providers that could support moving that capability to cloud. See examples in table 6.1.

What Are the Implications of Different Cloud Options?

While it may seem like things are straightforward when it comes to selecting as-a-service alternatives to fulfill your business and technical needs, it is not always that straightforward. Each choice creates not only a series of direct and immediate benefits, risks, and costs but also an enterprise-wide compounding effect when it comes to the scalability

Table 6.1
Cloud capability map

Capability	Category	Current model	Cloud model	Example cloud providers	Potential benefits	Potential drawbacks
Digital experience	Strategic	On-premise	IaaS	Amazon Web Services, Google Cloud, Microsoft Azure	Differentiation/ customization, agility, reliability, adaptability, scale up/down technology, optionality, availability of technical talent, partner ecosystem	Inability to differentiate/ customize technology, process standardization, complexity, vendor lock-in, managerial overhead, cost (w/o proper governance/ controls)
Accounts payable	Foundational	NA	SaaS	ERP	Process standardization, scale up/down technology, integration, availability of platform talent, configurability, partner ecosystem, reliability	Inability to differentiate/ customize technology and process, vendor lock-in, training costs, cost (w/o proper governance/ controls)
Pricing analytics	Strategic	NA	BPaaS	Deloitte, Vendavo, Vistex	Process standardization, scale up/down talent, scale up/down technology, process continuity, speed, value, capability, quality, reliability, access to data	Inability to differentiate/ customize technology and process, competitive differentiation
Cash management	Core	IaaS	BPaaS	Money center banks, fintech providers	Process standardization, scale up/down talent, scale up/down technology, value	Inability to customize technology and process, competitive differentiation
Other capabilities	...	...	...	...	...	...

of those choices. After you have identified candidate capabilities for moving to the cloud, and possible providers for those services, you need to think through the benefits and challenges for such a move. See table 6.1 for a list of common benefits and challenges associated with each. This list is not exhaustive but should serve as a starting point for discussion.

Take a moment to revisit your capability matrix, but now add in some additional dimensions to your capability documenting the implication of your choices—both for the technology and business capability. Does the value proposition and technical capability choice align? Is it different? What are some of the gaps? Are larger, more macro, capabilities required? Is there enough scale economics to justify multiple as-a-service models? Which make sense and where?

We acknowledge that none of this is easy and always recommend correlating these decisions with the people who have the right expertise in the right areas to ensure you are making the best decision possible for your organization. Show your work, talk to your colleagues and friends, and understand what is important to your organization and what is not.

7 How to Hyperdifferentiate with Data and AI

The big technology trend is to make systems intelligent and data is the raw material.
—Amod Malviya

When used correctly, data is not the enemy of Intuitive Creative Thinkers; it is a powerful friend.
—Leena Patel

For much of the past decade, we've heard about the promise of so-called big data to provide business insights. Data has been touted as valuable as anything from gold to oil for the modern business. The vast amounts of data that are created by our increasingly digitized world creates opportunities for understanding, automating, and offering new products and services within that world for those who understand how to leverage this new resource.

Furthermore, big data is inextricably linked to another set of technological advances—AI and machine learning (ML). Andrew Ng claims that "AI is the new electricity. . . . Just as electricity transformed almost everything 100 years ago, today I actually have a hard time thinking of an industry that I don't think AI will transform in the next several years." Throughout this chapter, we refer only to specific AI used for relatively narrow tasks rather than general AI that refers to a more human-like intellect that, for now, remains more in the realm of science fiction. Despite machines' ability to now compute faster than the human brain, they still aren't effective at mastering certain patterns or human capabilities.

Like many things involving digital technology, how data and AI provide value is often widely misunderstood. Furthermore, companies have evolved on how they are collecting and using data. Until recently, the prevailing wisdom was for companies to collect as much data as they could and figure out how to extract value from it later. Citing privacy concerns, customers have begun to push back on this indiscriminate collection of data, and many companies have shifted to being more purposeful about the type and volume of data they collect and store. Research has shown that customers often engage in a cost-benefit analysis when it comes to data privacy.[1] They are more amenable to their data being used if it actually provides them some value in terms of service or experience but are more resistant if companies collect it without providing them some value.

What's Different about Big Data? Volume, Velocity, Variety, and Value

The three Vs—volume, velocity, and variety—are often touted as the key factors that make big data different.[2] Yet we also add value to the list as the goal of the other three Vs.

- **Volume** represents the sheer amount of data being generated and stored on a near continual basis.
- **Velocity** represents the speed of data creation, which may be even more important for some companies and occur in real time.
- **Variety** represents the different types of data available, such as email, images, videos, chat, social media posts, purchases, medical records, location tracking, and connected devices.

Our addition of **value** to this list represents the need to build a mindset about how value can be extracted from data. We call this mindset, "data as a strategic asset," and it encapsulates the other three Vs. While value is not a characteristic like volume, velocity, or variety, it identifies the data that truly affects business strategy and outcomes and includes deciding what data not to collect because of customer privacy concerns or potential liability if that data is exposed. Executives with a growth mindset push their companies to answer value-focused questions about the advantages that can be gained through data. How can data fuel growth and innovation? How might data improve our culture? What datasets can I acquire or build that are distinct from my competition?

A great example of how to leverage data to address a crisis came in our discussion with Hitachi. Aligned with their principle of supporting sustainability, they used big data to help protect rainforests. Brad Surak described the work they did with Rainforest Connection, which involved collecting vast amounts (volume) of different types of mechanical and biological sounds (variety) in real time (velocity) to provide advance warning so that the appropriate authorities could engage in protection activities (value).

> We took old cell phones that can no longer run the latest software but are still totally fine as computers. We put them in a waterproof box with some solar panels on it, climbing up in a tree 150 feet and mount them in the rain forest. They listen 24/7 and connect to the cell phone network that happens to be in a lot of the rain forest. We stream that data to the cloud, and strip out the bio acoustic data to find human noises, like trucks and chainsaws. When the system detects these noises, it alerts the government or the Indian tribes that have jurisdiction, who can then intervene before the logging even gets established. If you put enough of them around the perimeter of a forest, you protect a huge swath. We then also provide the bioacoustic data to scientists to study the flora and fauna in a new way.

ML: The New Table Stakes for Data Analytics

The most exciting advances in data analytics are with respect to AI generally and ML, in particular. The terminology around these can be confusing. The AI terminology often conjures up dystopian science fiction examples like *2001: A Space Odyssey*, *Blade Runner*, the *Terminator* series, *Ex Machina*, and *Her*. In contrast, there are many different approaches to ML—broadly categorized as supervised learning, unsupervised learning, and reinforcement learning—that can quickly get confusing for the general reader. Throughout this chapter, we will use the term ML to try to strike a balance between the sensationalism often associated with the overly broad language of AI and the considerable nuance of the precise ML techniques employed for different purposes.

ML is an approach to developing information systems that relies on training the system on data rather than the traditional method of rule-based programming. With code-based systems, the programmers would have to consider and plan for any possible scenario the system would encounter, which made it very difficult to develop systems for complex and unstructured applications. With ML, by analyzing patterns in these large datasets, the system can learn what factors are associated with certain outcomes and then begin to optimize on those outcomes. The more and better data one has for training these ML systems, the better they will be able to perform their desired tasks. ML is helping advance fields such as image recognition, computer vision, natural language processing, and translation in ways that some computer scientists once thought impractical.

Companies are definitely beginning to take notice of the potential value of ML. The 2020 Deloitte Survey on AI in the enterprise notes that approximately 67 percent of companies have already used some version of ML in their organization, with some 97 percent indicating they expect to use it within the next year.[3] Furthermore, these companies have a plan for these experiments. According to the 2020 MIT SMR report on AI, nearly 60 percent say they have an ML strategy, up from 39 percent in 2018, and 87 percent believe that ML will allow their company to obtain or sustain a competitive advantage.[4] According to the 2020 Deloitte survey, 75 percent of respondents expect AI to transform their organization within three years, and 61 percent think it will transform their entire industry within that same timeframe.[5] Many of the ML platforms already have prebuilt algorithms. With the quickening adoption, if your organization is not using ML in any capacity today, it may be at a competitive disadvantage since there are already some commonly used models already available.[6]

One of our favorite examples of ML application is in the poultry management industry. It initially came as somewhat of a surprise to us to learn that chicken farmers are incentivized to produce chickens of a certain size for use in large-scale chain restaurants, such as KFC, Popeyes, or Bojangles. Farmers have long used software to monitor the growth of their chickens and make recommendations on feeding and other conditions to reach these benchmarks. One of these software companies experimented with ML methods to augment their recommendation software, and within six months the ML-based system outperformed the code-based software that the company had spent twenty years researching and developing.

This example illustrates the potential of ML for both creating disruption and weathering it as well. If a company could develop a disruptive ML system over the course of a few weeks, then their competitors who have access to similar data certainly could too. Conversely, if you have the data infrastructure in place, ML can be applied to disrupted environments in relatively short order. Recall the example in chapter 5 of Hitachi repurposing its factory sensor network to monitor for social distancing within two weeks.

Data and ML Doesn't Replace Human Intuition or Value

The most successful applications involve humans working with data and ML in various ways to obtain value. The 2020 MIT SMR report found that companies that can effectively find ways to allow humans and ML to learn from each other can significantly improve the financial benefit of ML applications. Thus, data and AI don't necessarily replace humans in the workforce, but humans work with data and ML in partnerships. Our colleagues at Deloitte have started to refer to this as the "Age of With," referring to the emerging partnerships between humans and ML.[7]

Some managers mistakenly assume that data and ML replaces human intuition, decision-making, and action. For instance, one large restaurant chain we worked with noted that its leaders had difficulty developing a data-driven culture because they had a sense that marketing was about "gut." Human intuition doesn't go away with a data-driven culture; it just means that you don't make decisions based on intuition alone. You can test those gut feelings and intuitive senses about what will be successful before you release them in the wild. The ability to test those intuitions in low-risk ways allows you to not only pursue more intuitive hunches but also sharpen and accelerate this intuition as you iterate through and learn from multiple scenarios.

The role of data in decision-making is similar to how data is used in digital design. Margaret Gould Stewart, VP of design at Facebook, says that "it would be irresponsible not to use data to tweak our designs. Data can help make a good design great, but it's never going to make a bad design good."[8] Data alone can't make decisions for your company, but a strong data infrastructure and culture can help make your leaders' decisions better. In fact, we think it is irresponsible for leaders to not take a data-driven approach to decision-making in this day and age. Furthermore, one danger of ML is that it can be difficult to know what it is learning from the training data. For example, when Amazon tried to use an ML system for filtering resumes, the algorithm exhibited a bias toward male candidates, simply because men had dominated hiring in the technology industry for most of the past decade.[9]

The importance of supervising or auditing ML systems is illustrated by Johnny Ayers, cofounder and senior VP of the identity verification and fraud protection company Socure. Socure's target market, financial services companies, have robust risk systems in place to identify biases in the way they offer their products to the public. (Lending bias is closely regulated, specifically with Regulation B from the Equal Credit Opportunity Act from the Consumer Financial Protection Bureau.) Socure is extraordinarily careful to ensure that its credit reviewing and granting models don't contain biases related to "age and gender and race and socioeconomic status," Ayers says. As a result, the company had to be equally mindful of these issues as it created its services. Ayers adds that "even when we were only 10 people, we were building a lot of very specific controls into how we build and train models, knowing that, when you sit down with any number of the major credit issuers, their expectation is that you have stress-tested any of your models that you're proposing to ensure that none of the aforementioned biases are implicit in your models."

In many ways, human-machine partnerships are already widespread in business and society. Our mobile devices already virtually eliminate the need for certain types of information to be stored in long-term memory (e.g., phone numbers) and reduce the importance of certain types of intelligence (e.g., spatial awareness through Google Maps, time and task management through calendar apps, social intelligence through online networks). For that reason, we refer to these human-machine partnerships as ushering in an "Age of With," where most human work is conducted in conjunction with computerized support.

Although the specific applications of data and ML are virtually limitless, they can be grouped into three broad categories:

- With **automation**, ML uses data to allow information systems to perform routine actions on behalf of human workers. For example, many auditing tasks are now being performed by ML systems. Automation shifts the value humans bring to the organization to different types of roles.

- With **augmentation**, ML partners with humans at the individual level. In this setting, ML provides insights that humans can use in their decision-making. It's been said in the healthcare industry that ML will not replace doctors, but doctors who use ML will replace doctors who do not.

- With **application**, the organization is using data and AI to develop new products and services. Kevin Kelley, founder of *Wired* magazine, comments that the next wave of start-ups will simply take an existing problem and apply ML to it, like the chicken farmer example we discussed earlier. We would caution that any application of ML also needs to involve deep domain expertise to achieve value.

Building a Data-Driven Culture

Understanding how data can provide value to a company is key to deriving value from it. Gartner estimates that 60–85 percent of big data projects fail to deliver their anticipated results. That's because many companies don't know where the value of data comes from. Yet, too often, companies build the infrastructure to collect and analyze massive amounts of data without considering what questions they will answer with it or whether the answers to those questions are worth the amount of investment they are placing in it. Pablo Picasso famously noted "Computers are useless. They can only give you answers." Data has the same uselessness as computers, but if managers can ask the right questions and learn to derive value from a robust data infrastructure, the outcomes can be powerful.

One of the reasons data initiatives often fail is because the organizations lack a data-driven culture. Building a data-driven culture that can derive value from this abundant data, however, is not easy. A 2019 Deloitte survey showed that most executives don't believe that their companies are data driven, and 67 percent say they are not comfortable accessing or using data from their existing tools and resources. Somewhat alarmingly, the percentage of respondents who say their company is data driven has actually decreased in recent years, from 37 percent in 2017 to 31 percent in 2019.[10]

Yet, this data-driven culture can be difficult to cultivate in legacy organizations. Tom O'Toole of Northwestern University says, "You have to accept that this is very uncomfortable. A data-driven approach to the business inevitably introduces transparency, there's no getting around it. That's not welcome in a lot of cultures." O'Toole tells the story of a young PhD data analyst at a financial services firm. The analyst was tasked with identifying which retail bank branches were not profitable and should be closed. Upon submitting his analysis to management, the analyst was told by the executive in charge of retail branches, essentially "Your analysis is never going to see the light of day. I'm going to be sure it goes nowhere. I am in charge of retail branches, and I don't believe in closing retail branches." Once leaders establish that unwelcome answers from data analysis are not tolerated, it can rapidly kill a data-driven culture.

Brice Challamel of Google Cloud refers to a data culture as the first step on the journey to data mastery.

> First, companies need to learn to make decisions based on data. It's totally unrelated to technology, it's a matter of culture. The second step is data consolidation. You need to organize and curate data, so that it doesn't remain in different silos disconnected from each other. Third, you need to be very proficient at analytics, which is how you use that data to generate a clear understanding of the current environment for decision-making. Fourth, how can you leverage those analytics to train models? Analytics are the present, but models are perspectives regarding the future that enable you to predict, recommend, and optimize for the longer term. Lastly, you move toward data mastery. Data mastery is where you have a perfect fabric of data workflows that connect to each other across your value chain, feel each other, support each other's accuracy, sustainability, and capacity to be accurate and on point in a timely fashion.

Developing a data-driven culture requires a certain amount of education and experience that employees cannot get in their day-to-day work. Thomas H. Davenport describes one effort by TD Wealth, the wealth management unit of Toronto-based TD Bank Group, who sought to develop a more data-driven culture. It launched a program called WealthACT that sought to accelerate the company's digital and data culture. The participants were competitively admitted to the program and involved formal educational components as well as visits to, and working sections with, data-driven companies in Boston, Silicon Valley, Israel, and Montreal. It allowed participants to develop the formal knowledge they needed to develop analytics literacy as well as the practical experience to know how to use it effectively.

Data Enables Nimbleness through Sensing and Responding

Data enables organizational nimbleness by increasing an organization's capability to sense and respond to environmental changes. It does so by enabling rapid experimentation, enabling real-time tracking of environmental conditions, and by applying existing data capabilities to new situations.

One way that data enables nimbleness is that it turns the organization into a massive "natural experiment." A natural experiment is one in which you study differences in responses to events that are outside the researcher's control. With respect to acute disruption, it allows you to track the different way people respond to disruption and analyze these responses to identify which are most successful. Ben Waber of Humanyze notes the importance of data-driven experimentation, saying that "it's an unfortunate natural experiment. Many of our clients have tens or hundreds of thousands of employees globally. They're trying a number of things to try to improve collaboration across the board remotely, but without analytics it's very hard to know what actually works."

This ability to use data to turn disruption into natural experiments is particularly valuable for global companies, who may experience an acute disruption in different ways and on different timelines. Waber continues, "Our customers are looking at East Asia—China, South Korea, and Japan to a lesser extent—as a test bed for some of those ideas. Using these experiments in parts of the world that are further along with respect to their experience of the pandemic to help drive an understanding of what appears to work and what doesn't within their organization." Many of our respondents at global organizations pointed to it as a strength in responding for just this reason.

McDonald's also used data to enable nimbleness for sensing and responding quickly. For example, they sought to employ digital kiosks to reduce congestion at the front counter and adapt to customer preferences, as well as improve throughput in restaurants. Chris Whitfield explains the results, saying "In the US, kiosk usage among our customers was low, and they were all just going to the front counter. In almost every other market, the opposite was happening. We were at risk of essentially shifting resources away from a place that was a source of growth in a number of different markets." Data allowed McDonald's to be nimble and tailor their strategic actions to particular markets.

Data also enables nimbleness by using the increased velocity of data to make decisions faster than would be possible otherwise. For example, the location-services firm Foursquare partnered with the city of New

York to use real-time location data to monitor people's activities and identify lower density times for shopping, allowing people to adjust their behavior in response.[11] Likewise, Steelcase used relatively simple data visualizations as a tool to track how lockdown orders in different states and countries would affect both their upstream and downstream supply chain. Sara Armbruster explains, "It was so fluid, where things were changing day by day. Our data visualization team created this awesome map of the US. They had every single one of our distribution centers, facilities, and dealer partners on it. They created a red, yellow, green coding that indicated each facility's status at the moment that was so helpful for making decisions." The data visualization allowed operational leaders to make better decisions in real-time, enabling the organization to nimbly respond to changes on the ground as they occurred.

Lastly, data enables nimbleness, because it can be relatively easy to bring an existing data infrastructure and culture to bear on a new problem. Peter Schardt, CTO of Siemens Healthineers, notes how they were able to apply their investments in ML to help analyze new data about COVID. He describes the use of ML in one of their medical scanners, saying, "Because we had invested in these Machine Learning capabilities, it's been relatively easy to add COVID to the list of abnormalities we scan for and track, which has helped us develop a very valuable diagnostic tool for COVID in relatively short order."

While the above examples are all from addressing the specific COVID disruption, it is relatively easy to see how a robust data infrastructure could be applied to provide nimbleness to respond to disruptions of other types as well. For example, Humanyze's experimentation infrastructure could be applied to assess a number of changes in the business environment, including those implemented by the organization itself. For example, several years ago in foundational research, Humanyze worked with one large technology firm to determine that changing the size of the lunch tables would result in a greater than 10 percent increase in employee productivity, fostering greater communication and cohesion among the employees that led to improved knowledge sharing.

Data-enabled nimbleness is also valuable for other types of acute crises. For example, the American Red Cross established a social media listening system to scan for keywords such as "tornado" or "earthquake," allowing them to provide information resources more quickly to people experiencing a natural disaster.

Data Enables Scalability by Making Large Datasets Actionable

Data enables scalability by allowing organizations to process vast amounts of data and make better decisions regarding large swaths of customers, employees, or products at once. With the increasing volume of data being created in today's environment, it can often be difficult to assess that data without a strong data infrastructure and culture. ML is a particularly valuable tool for this purpose because it allows you to make sense of large sets of data in ways that may be difficult otherwise.

For example, Rajeev Ronanki of Anthem explained how they used ML during the Regroup phase of acute disruption. Specifically, it allowed the organization to analyze their consumer base to identify which individuals they should encourage to seek care for other conditions once the initial COVID surge had passed.

> We knew consumers were putting off necessary care. We identified these people through AI and our predictive algorithms. We have claims data for over 70 million people, clinical data for over 10 million, and lab data for over 15 million members. With that type of data, AI can identify the people to whom our intervention should be targeted. Rather than sending it to several hundreds of thousands of likely candidates, we were able to narrow the scope of our outreach to just a few that needed it most critically.

Another example of scalability can come from medical research involving COVID. With so many of the world's research labs focusing on various aspects of COVID, a massive number of results were generated—28,000 research papers as of June 2020. That volume makes it difficult for individual researchers to identify relevant findings that could propel their own research forward. Researchers developed a ML tool that helped these individual researchers identify which of these papers were most relevant for their own research, sometimes highlighting specific findings. Beyond the pandemic, the developers of the tool noted that similar tools can help bridge academic discipline by identifying relevant research that may be occurring in other fields.[12]

Data Enables Stability through Improving Decision-Making

Perhaps the most important capability of data is stability, because it allows leaders to make decisions based on data rather than simply rely on intuition or guesswork. Erik Brynjolfsson and Andrew McAfee note that this process allows organizations to move beyond relying primarily on HIPPOs—the "highest paid person's opinion." This stability can be particularly important during times of disruption, when a manager's instinct or experience may not apply to the disrupted environment. For example, Sara Armbuster of Steelcase described how they were using data to ensure that social distancing occurred in the workspace.

> All of our spaces are outfitted with sensors for tracking occupancy and utilization of facilities. Our facilities and real estate team looked at the data generated by this sensor network to identify areas of high historical utilization, occupancy, and density. We could track utilization minute by minute and hour by hour, creating fairly complex heat maps to identify hotspots. We adjusted the spaces or schedules to alleviate the problem before employees returned to the office.

Data can provide stability in other types of disruptions as well. For instance, an initial trigger that led the retailer Bed Bath & Beyond to move toward a more robust data infrastructure and culture was the yearly disruption of Black Friday and Cyber Monday, the two biggest shopping days in the United States. Inspired by the success, they have invested more in their data capabilities to enhance the company's abilities to project future sales trends, make better decisions based on real-time customer data, and improve demand prediction and inventory fulfillment— capabilities that increase the stability of the business operations.

Doug Mack of Fanatics, an online retailer of licensed sportswear for major professional sports leagues, described a similar application of data to manage disruption: "We're a highly analytical business that has tons of historical data demand curves. Effectively, we have a forecast by our major sites of what demand is going to be. That kind of central nervous system gets ripped away from you when you have such a macro shock where the pandemic moves sports off the field." Fanatics had to rebuild those data capabilities for the new environment that forecasted demand in a new environment while also performing "a constant health check on our ability to manage and sell inventory as it came in, while making sure our fulfillment and call centers are safe." He describes how this data supports decision-making on a daily basis: "Do you lean in, or pull back on different sources of marketing based off of the available consumers out there and the cost to acquire them? Are there certain types of promotions that are more effective, like free shipping? What is the optimal conversion behavior on a day in day out basis? So, we were doing a lot of testing."

Data Enables Optionality through Interoperability

Optionality with respect to data mostly involves interoperability that allows different datasets to be combined that provides greater insights that either one could manage on its own. For instance, Brad Surak, formerly of Hitachi Vantara, describes a partnership with the American Heart Association that involved combining datasets from different research efforts to create what they called the Precision Medicine Platform. This platform allowed different research teams to upload their research data, which was then "harmonized" into a standard format to allow interoperability between the datasets. Researchers could then combine various datasets to ask a wider array of questions and to perform robust analysis using existing datasets rather than having to continually collect new ones. Surak says that "it was about getting data agility. How do we go and enhance the sharing of data around cardiovascular disease amongst the research community, so that we can more easily propagate these data sets and we can help these researchers collaborate and identify paths to treatments more quickly?" Once this data sharing platform was in place, the American Heart Association could easily shift it to support COVID research. Surak explains that "the American Heart Association was able to take their expertise and leverage it to convene researchers around COVID rather than heart disease."

Krishna Cheriath, head of digital, data, and analytics at Zoetis, notes the benefits of the optionality created by the interoperability of data in the healthcare industry.

> One of the biggest challenges we have is the interoperability of data in the healthcare ecosystem. If healthcare data could move in a frictionless manner, it would allow for innovations to happen at much greater speed and scale. Is it 2×, 3×? I don't know, but I'm convinced that, as a society, if we had a better-connected data ecosystem, we could respond much more quickly. Clinical trials and clinical development can go much faster.

How Does Data Help with Disruption?

Data can help companies in every stage of adapting to disruption. In the Respond phrase, it can help by letting leaders get a better sense of how things have changed. Whether it be by analyzing the effect of remote work on employee productivity, by providing greater transparency into supply chain disruptions, or by analyzing changing customer demand patterns, data provides the clearest view into exactly how an acute disruption has affected business operations. This ability to collect, monitor, and analyze data on a changing environment provides the necessary support to help companies respond effectively by keeping business operations functioning during disruption.

Data can also help in the Regroup phase by helping identify trends that companies can begin to capitalize on. If leaders have a data-driven view into the types of behavior shifts that are occurring in the disrupted environment, they can know when to begin restoring services or what new types of goods or services may be in demand in the marketplace. For example, the clothing retailer Zara has long monitored what shoppers are looking for that the company doesn't currently carry. This capability would allow Zara to identify demand changes and to produce goods shoppers are interested in.

Last, data can help in the Thrive phase by providing a robust experimentation architecture that allows companies to be continually adapting to a changing world. Put simply, we think data capabilities and culture were going to be essential for companies thriving in a digital world regardless of acute disruption. Hopefully, this crisis has simply illustrated more clearly to most leaders the value of these capabilities so that they will begin to develop them at the individual and organizational level.

Brice Challamel of Google Cloud describes the significance of which data can change the business environment.

> Do verticals still matter? I believe they don't anymore. In the age of data, It's not about verticals. It's about business archetypes, and how technology can support their evolution. There are more similarities between a gas station and a postal office—let's call them distributed networks of services—than between that same gas station and an oil rig in the same vertical. There is more similarity between Robinhood and Chico's—communities of enthusiasts—than there is between trading and life insurance in the financial services vertical. Verticals are dictated by the legacy constraints we inherited from the industrial age and they are failing us in our effort to understand and standardize technology-driven transformation patterns.

How to Apply the Concepts from This Chapter: Questions for Your Data Ambitions

As we have noted, data is necessary but not sufficient to generate organizational value. It is the thoughtful use of data that allows the organization to identify and drive actionable insights. We have crafted an activity that will help you to begin to understand which data assets are available, which are valuable, and which choices make the most sense—in the context of further advancing business capabilities and outcomes.

What Are My Organization's Data Ambitions?

Before you can begin down the path of advancing business capabilities or forming new revenue streams, you must first define your data ambition. Does your organization intend to incorporate data into everything you do? Do you want to be a net producer of data? Do you need data to better improve the quality of experience with products and services? Is data primarily a resource to improve business capabilities and processes? Is data owned or shared? What advanced data capabilities are applicable to your business (i.e., deep learning, ML AI)?

Take a few minutes to write down your data ambition. If you are looking for some inspiration, see below for ideas.

- Enable stability by building systematic intelligence into everything you do or gain deeper understanding into business via data science.
- Facilitate scalability by democratizing institutional insights and learning through greater access to data across the organization.
- Create nimbleness by leveraging data to launch new products and services or to improve and augment existing ones.
- Provide optionality by making your data infrastructure compatible with partners or by participating in a data platform where partners participate.

What Foundation Is Needed to Achieve Your Data Ambition?

Now that you have a good idea of your overall ambition, let's think about what's needed to enable that ambition. The foundation needed will depend heavily on the nature and extent of your ambition. Do you have the talent to deliver your ambition? Do you have a sufficiently progressive data culture? Do you have the technology? What policies and procedures do you need to protect privacy? Confidentiality? What processes do you need to make data available for use? Do you have data governance policies and procedures? Do you have any new business models for data monetization?

Take a few minutes to write down the key components of your future data foundation. See below for some additional inspiration.

- **People and culture**: Improve experimentation and testing, enable people analytics, balance data and experience, be open to data-driven results, infuse the organization with data-native talent
- **Data modernization**: Ingest, translate, move, connect, and store a great range of data; expel unnecessary and nonuseful data

- **Applied analytics**: Visualize and share insights more consistently to achieve one version of the truth, predict and model future potential outcomes to a broader set of functional areas

- **Automation**: Standardize and automate repeatable processes, experiment with ML on more complex interactive processes

- **Governance**: Establish clear roles and responsibilities for data ownership and upkeep, develop a top-down point of view on data privacy and AI ethics

Do You Have the Data Needed to Meet Your Ambitions?

Last, before getting started you should take stock of the data balance sheet. Take a few minutes to write down your data assets and liabilities. If you need some inspiration, see below.

- **Data assets**: What unique data does your business generate? What data does it acquire? How can existing data be combined to form proprietary data or insight? How is data being stored so that it is readily accessible for insight, action, and automation?

- **Data liabilities**: What data does your organization lack? What data do competitors generate that provides unique advantages? What market actions are constrained by a lack of insight or data?

If you did everything right, you should now have a better understanding of your data ambition, data foundation, and data balance sheet. If you are feeling a little overwhelmed by the enormity of this topic, we get it—data mastery can be daunting. If you completed this exercise, you should now have a sizable list to get the ball rolling with your technical and business colleagues.

8   How to Ensure a Cybersafe Future

How many IoT devices exist, with how many computing devices do they share data? How many others have access to that data and what decisions are being made with this data? No one really knows. We just don't know.
—Rebecca Herold

While we generally view the accelerating digitalization of our economy and organizations in the face of acute disruption to be one of the silver linings of the acute COVID disruption, providing the opportunity to make changes to modernize operations that have been a long time coming, it also introduces new potential risks. The proliferation of technology and extension of an enterprise's boundaries requires new approaches and methods for securing it. Historically, enterprises have just set up data centers and locked them down like Fort Knox. With the rise of the cloud and other distributed computing approaches, however, security must adapt and become more distributed. The old model of putting all your eggs in one basket and watching it won't suffice.

As companies virtualize more and more of their enterprise—work and collaboration among employees and partners, customer interactions, transactions with suppliers—risks to their operations increase if their data or operations become compromised. The individuals supporting and managing technology are more distributed and systems more complex, and the level of sophistication and volume of bad actors and threats are ever increasing.

For example, AI will likely be increasingly used for both attacking and defending organizations' digital infrastructure. We've already identified the benefits associated with increasingly automated and connected factories, yet they also come with an equal or greater number of associated cybersecurity risks as well. The same applies to the growing field of autonomous vehicles and connected devices. Each of these new devices and applications provides both an immense opportunity but also a potential weakness to exploit.

The challenges associated with cybersecurity are going to get more complex and potentially into unchartered territories as technology and the associated business model and operations propel us in the future. Brad Surak, formerly of Hitachi Vantara, describes these future business opportunities.

> The more exciting part of digital transformation is the convergence of hardware and software and the Internet of Things in our world. You can start thinking about your traditional products as a channel for digital revenue. When I put that into the industrial businesses, if an industrial air compressor can be made into a connected product, it becomes a channel for business model disruption. I can sell air in an operational expense model, rather than selling machines in a capital expense model—we can also do the same thing with a train or almost anything else. The use cases are there and they're very cool.

We completely agree that these new connected devices represent an enormous business opportunity. At the same time, they also represent a potential weakness for your company as well, because each one then becomes a point of vulnerability to hackers. If hackers gain access to this equipment, they could shut down the connected air compressor or an entire factory until you meet their demands, a rapidly growing phenomenon called ransomware. If you do not meet their demands, they could cause the machinery to malfunction and cause damage. These problems are not possible with more traditional equipment that is not connected to the internet. Thus, we felt it important to include a chapter on cybersecurity so that leaders begin thinking about these security implications as they engage in rapid digital transformation.

This Is a Different Type of Chapter

The issue of cybersecurity is becoming so complex that we found it impossible to cover all of these issues comprehensively in a way that would not be obsolete by the time the book is actually published. Each new security expert we spoke with recommended an entirely new list of challenges they believed was imperative for us to address. For example, one person helpfully pointed us to the MITRE ATT&CK framework as a place to start,[1] a framework that involves hundreds of different enterprise security considerations across eleven different categories. We simply cannot possibly cover that level of detail here. But just because the problem is complex doesn't mean it can be ignored. The complexity led us to take a slightly different approach to this chapter.

First, this chapter is somewhat different than the others in this section. While cloud computing and data/ML enables nimbleness, scalability, stability, and optionality, robust security primarily provides stability. Yet, we argue that this stability merits a separate chapter because what good is it to develop nimbleness, scalability, and optionality if it makes your company's operations vulnerable? The stability provided by cybersecurity not only drives down risk but can also create a business advantage through engendering trust among customers, partners, and employees that drives real value for the company. For example, Apple is one leading company that is emphasizing trust, security, and privacy as a key selling point for its products.[2] Andrew Burt, chief legal officer at the data governance firm Immuta, emphasizes that cybersecurity is really about putting customer trust at the center of competition.[3]

Second, unable to address all of the necessary technological aspects in a single chapter, we focus instead on the cultural, organizational, and people aspects of cybersecurity. Lisa Rager, head of risk management at Tesla, describes this approach to security. She said, "We take a people-focused view of security. We focus a lot on training and awareness. Lack of awareness and lack of understanding about the appropriate way to do things is the number one cause of data loss. Technical tools don't train your workforce to know how to do things the right way."

This organizational approach doesn't mean that the technological aspects are somehow unimportant, but rather we are focusing on the aspects of cybersecurity that may be under the control and experience of the average manager. Effective cybersecurity is everyone's job, not just for security professionals. As a result, some have argued that the best cyber-security investment you can make is training employees better.[4]

Third, this chapter is focused less on trying to guide you on *how* you should engage in effective cybersecurity and more on convincing and inspiring you that you *should* take cybersecurity seriously. For example, although some may see it as good news that 49 percent of respondents to Deloitte's 2019 cybersecurity survey indicated that cybersecurity issues are on their board's agenda at least once per quarter, it can also be interpreted as over half are not discussing security issues as often as they should. Cybersecurity should be one of the top risks boards should be having on their minds. It's not about if breach will happen but when, and boards need to discuss how to handle it in advance. Further, the implications of these discussions need to filter down to the actions of your employees. If your organization is not having regular discussions about cybersecurity at all levels of the organization, start now!

The Importance of Cybersecurity during Acute Disruption

The issue of cyber security is pressing with respect to disruption. The acceleration of digital transformation that we observed in many companies also makes companies more susceptible to various types of adverse digital events.[5] Our concern is that with the speed that many organizations are transforming, leaders are only considering the opportunities that come with the shift and are not considering the risks that also may be associated with it. In fact, our research on chronic digital disruption showed that executives were far more likely to consider the opportunities that digital transformation presented while often overlooking potential threats. While we have emphasized many of the benefits that companies have experienced with accelerating digital transformation in this book, we should also acknowledge the possible threats.

Cyberattacks and cybersecurity was already a major issue facing most organizations in a world of chronic disruption. Respondents to Deloitte's 2019 cybersecurity survey showed that 95 percent indicated that their organizations had experienced a wide range of cyberattacks, with 57 percent indicating that their most recent cyber breach happened within the last two years. Among the biggest impact of these breaches are loss of revenue due to operational disruption (21 percent), loss of customer trust (21 percent), change in leadership (17 percent), reputational loss (16 percent), regulatory fines (14 percent), and drop in share price (12 percent). Additional impacts may be experienced for customers, such as identities being stolen, financial fraud, and account hijacking.

For three years running, survey respondents identified that chronic digital disruption—rapid IT changes and rising complexities—is the most significant cybersecurity challenge organizations face. The problem of cybersecurity, however, has only become more challenging amid disruption. Regardless of the source, threats are possible twenty-four hours a day, seven days a week, including when—and especially when—you are at your most vulnerable. When employees are stressed, they may be more likely to cut corners or drop their guard. Bad actors also know that a time of weakness is an opportunity that can be exploited. Furthermore, the acute COVID disruption challenged many existing approaches to cybersecurity. For example, many organizations' cybersecurity infrastructures weren't built with remote work in mind, and many organizations had to make key risk decisions when faced with lockdowns, giving up on certain cyber controls to keep employees productive.

Cyberthreats are likely to accelerate in the face of acute disruption, and we witnessed just such an uptick in the wake of COVID. Cyberattacks increased by approximately 33 percent.[6] COVID was increasingly used as a lure in phishing attacks,[7] and the World Health Organization experienced a fivefold increase in attacks.[8] A large IT organization was the victim of a ransomware attack in April 2020 that resulted in $50 to $70 million in damages.[9] US research institutes experienced cyberattacks where data about vaccine research was stolen.[10] Schools forced into remote learning settings using the video platform Zoom began experiencing

"Zoombombing" attacks, where unauthorized outsiders would disrupt online classes using hate speech or pornography.[11] COVID played a significant factor in social engineering, in which attackers use interpersonal deceit to gain access to systems or data.[12]

One of our interviewees described healthcare providers having to protect against attempted attack in the midst of the acute COVID-19 disruption. Kristin Darby of Envision Healthcare explained, "The threat was basically an exploit that they were using to change routing numbers for accounts payable payments to have distributions rerouted. They replaced the routing number with their offshore checking account. You think you're paying your bills, but the money goes instead to an offshore bank account." These examples simply serve to show how attackers will use disruption as an opportunity for attack, adapting their methods to whatever the disruption may be. Rajeev Ronanki of Anthem notes the correlation between digitalization and the need for cybersecurity: "As data becomes more important in the delivery of care then the privacy and security of that data continues to be top of mind. That was the case before and, if anything, it's even more important now as we deliver care services virtually and we use more and more digital solutions in the larger ecosystem."

Furthermore, cyberattacks can be a risk, even if your company is not the intended target. For example, the 2017 "NotPetya" cyberattack resulted from a Russian intelligence service launching a cyberattack on neighboring Ukraine that escaped the bounds of their target.[13] It eventually spread to infect companies like Merck ($840 million in estimated damages), FedEx ($400 million), Maersk ($300 million), WPP, and many more, becoming the most destructive and costly cyberattack in history. Complicating matters are the developing legal questions regarding whether existing cybersecurity insurance policies will cover many of these losses since the attack has been categorized as an "act of war," which is not covered under the policies. As of this writing, the legal resolution is still being disputed in the courts. As acute disruption increases the likelihood of cyberattack, it also increases the chance that your company will be collateral damage for an attack on someone else. The number of state-based cyberattacks has increased during the acute

COVID disruption, which could make an increasing number of companies into collateral damage in these attacks.[14]

Since many organizations report that they are unlikely to go back to the "old way" for the foreseeable future, the challenge of cybersecurity is here to stay, and digitally resilient organizations must tackle these challenges head-on. We have some evidence that organizations are doing so. Mark Onisk of Skillsoft notes that their online materials related to security operations had increased dramatically during the crisis. In much the same way that acute disruption can be an accelerant for digital transformation, it can also be used as an opportunity to improve the organization's security. If organizations keep security in mind, they may end up in a more secure setting, that is, if security has been built into these processes as they've developed them—adopting "security by design" principles during the development and implementation of these new technologies to manage disruption.

Rajeev Ronanki of Anthem echoes these points.

> The number of points that need to be protected have vastly increased as telemedicine and digital means of interaction and data are made available to multiple stakeholders. We can't just rush to create solutions that meet a functional need. The hard work is in creating the underlying infrastructure that's needed in order to protect privacy and security and keep the trust in that entire interaction.

Building a Culture of Reliability

Cybersecurity can be a challenging problem because it most frequently comes from within your own walls.[15] A breach is not always something that is a result of an actual "attack." Cybersecurity covers a broad set of topics, and one of the most common forms of breach is simply due to employee error. The actions of well-meaning employees are actually a bigger security threat than technical vulnerabilities. The 2020 Verizon Data Breach Investigations Report found that phishing attacks, where employees are tricked into clicking on a malicious link, are the cyberattacks most likely to lead to a data breach.[16] One cybersecurity expert we spoke with indicated that a key reason people often click on phishing links is they just get overwhelmed with the email load and don't approach it as carefully as they should.

No matter how secure your IT systems are, that security is compromised if your employees unwittingly let those attackers bypass it. Indeed, many of the most challenging aspects of cybersecurity management are not technological but organizational. Krishna Cheriath of Zoetis echoed the importance of this security-centric culture in terms of digital citizenship.

> Companies should cultivate the notion of digital citizenship. Every employee has a role to play. They need to have a good awareness around basic things like cyber vigilance. They need to be data responsible and make sure that they're responsible for ethical consumption of data. How do you make sure that you're protecting the company's assets by making sure the appropriate actions from a cyber or data perspective?

One might look to the US military as an example for the type of integrated technological and cultural approach for cybersecurity when dealing with the chronic digital disruption.[17] Knowing that it was likely to be a significant target for cyberattacks, in 2009 the Pentagon engaged in a massive effort to make its operations more secure. It believed that a technology-first approach would be unlikely to be successful, so it borrowed concepts used decades earlier to secure the eighty-three nuclear powered vessels in its naval fleet to create a high-reliability organization. They focused on developing six principles:

1. Integrity: The first and foremost objective is to create a culture where employees understand the consequences of security lapses but also encourages employees to speak up if a mistake has been made.
2. Questioning attitude: Not only are employees expected to double and triple check their work, but they are also expected to question their environment and "say something" when something seems amiss.
3. Depth of knowledge: Employees need to understand the whole operation of the organization, enabling them to better identify when something seems amiss.

4. Procedural compliance: Employees are expected to know and follow the rules to the letter. Extensive inspections occur to ensure these rules are being followed.

5. Redundancy: Activities are carefully monitored in real time, with high-risk activities requiring that two people perform them.

6. Formality in communication: To minimize the risk that misunderstandings occur, instructions are given clearly and then repeated back verbatim to be sure they are correctly understood.

This approach appears to have been successful. Between June 2014 and September 2015, the US military repelled thirty million attempted cyberattacks, with fewer than 0.1 percent having any meaningful effect on the systems or operations. Although many of these principles would clearly need to be adapted to a civilian context, they do serve to demonstrate ways in which effective cybersecurity needs to go beyond simple technological solutions. It is important to strengthen the culture of your organization to reduce the number of "weak links" that can be exploited by cyberattackers or simply due to user error.

Rethinking Cybersecurity Leadership

One challenge many companies face with respect to developing effective cybersecurity is leadership. Just because you have a chief information security officer (CISO) doesn't mean security is all that person's responsibility. In this chapter, we emphasize that cybersecurity is everyone's job in the organization. Cybersecurity should be embedded in each and every business decision. For example, when a business decides to put something on the public cloud, the onus is on the business to protect it. The concept of shared responsibility is an important aspect to handling cybersecurity, and the fact that it's shared and everywhere makes it so much more challenging. Tom O'Toole of Northwestern University notes the importance of cybersecurity across the entire organization.

Does the increased use of data create any additional cyber security risks? Enormously so. Few things have the potential to destroy all of the great advances we have achieved faster than cyber security breaches. Even people in functions that are not directly responsible for cybersecurity have an enormous vested interest in sound cybersecurity. I'm not intending to sound overly dramatic; I really mean it.

That said, it is also important to have strong leadership who does have a comprehensive view of an organization's defenses and to guide these efforts. If cybersecurity is an issue that crosses the entire organization, who should be the CISO and where should that person fit into the organizational hierarchy? Because cyber risk extends beyond the enterprise systems to all the products and services delivered by the enterprise, it doesn't make sense to have it entirely bound within the IT organization. Reporting directly to the CEO is the single most commonly reported position for the CISOs, yet it still only represents less than a third of companies.

Nevertheless, the position is often situated within the IT function. Approximately the same number of respondents indicated that the CISO reported to the CIO or CTO. There can be benefits to this approach if it allows companies to bake security into their development and implementation efforts. On the other hand, if these leaders primarily come from a technological background, it can lead to ineffective security leadership cultural dimensions being ignored in favor of a technology-centric approach. As George Westerman, principal research scientist with the MIT Sloan Initiative on the Digital Economy, noted, "It's never been a better time to be a great CIO or CTO, but it's never been a worse time to be an average one." We consider great CIOs or CTOs as those who exhibit both expertise in the technical and business/organizational aspects of technology, and a CISO needs this combination of business and technical knowledge as well. It may be okay for a CISO to report to a great CIO but perhaps not to an average one.

Protecting the Crown Jewels

You should also assume that—despite your best efforts—you will experience a data breach at some point in your future, and you should develop various plans for how to react when you do.[18] A data breach can also be considered an acute disruption, but it's one you can antici-pate building resilience to in advance. Eric Ranta of Google Cloud emphasized the ubiquity of cybersecurity threats: "The recent example of a 17-year-old from Tampa hacking into Twitter's platform shows you that nobody's really all that safe. Twitter is as sophisticated as anybody, and mainstream companies—whether it's retailers, healthcare, or oil and gas producers—are all subject to it."

The first step to building resilience is to create a security event or major incident response plan. The plan should include a current list of data resources, a description of those resources' value, and the implications of that breach. It should consider a list of potential threat actors for your organization as well as potential responses for each type of attack. It should also include a list of whom to contact in the event of a particular type of attack and when, including multiple channels by which those parties can be contacted if one of those channels are compromised in an attack.

It's difficult, if not impossible, to protect all of your data assets com-pletely. As a result, companies should perform a strategic data assessment, focusing on identifying and protecting the most valuable assets and opera-tions.[19] The core of the enterprise requires the strongest defenses, whereas the periphery and limbs are more expendable. After completing a value assessment of your various data sources, invest in security relative to its strategic value. This value assessment not only considers the upside value of data, such as proprietary information used to enable your organiza-tion's critical success factors, but also the downside value of data, such as regulatory penalties if that data is somehow breached, particularly if laws were broken or controls violated or left unmitigated. This process should be repeated periodically to ensure that knowledge is regularly updated.

Engage in Cyber Wargames to Build Resilience

Once this plan is in place, you should periodically engage in cyberse-curity "wargaming," where you test these security practices. Through these tests, you learn and improve your plan and also drive appropri-ate investments to build the resilience capabilities. While you probably can't effectively plan for the exact cybersecurity incident you will face, you can build the muscle memory of effective responses to develop resilience when the breach eventually does come. This incident response plan should be regularly tested against these wargaming exercises and then reviewed and revised in response to what you learned during these tests.

Kristin Darby of Envision Healthcare describes the company's cybersecurity planning efforts.

> Prior to the pandemic, my cybersecurity team practiced scenario-based exercises on a monthly basis and more in-depth scenario exercises quarterly as well as running incident command activities. Our cyber planning is probably the most mature process that I have experienced for scenarios based on actual events, and our ability to quickly respond to the COVID-19 crisis was certainly a benefit.

Furthermore, the nature of threats may change, and security leaders need to remain aware of this changing landscape. Deloitte, Verizon, and other companies regularly publish briefings on current cyber threats. Security executives should regularly review these as well as provide high-level reports on these trends to management. Even the simple process of reporting on the changing threat landscape to the board and senior executives can help ensure security concerns and awareness remain front and center in their strategic thinking rather than be applied as an afterthought.

This process should also involve explicit efforts to improve your organization's cyber hygiene. In addition to user error, software and malware are also the two most popular breach event drivers. Between one-quarter and one-third of all breaches involve known vulnerabilities that simply go unaddressed. Ensuring that the security protocols of all systems are up-to-date essential for enterprise security. Yet, patching itself can also create new vulnerabilities. Companies should evaluate and test their patch levels before deploying patches and do so promptly to make that known vulnerabilities are quickly addressed.

Unfortunately, our data shows that not enough companies are engaging in this type of planning or review. Less than half of respondents indicate that they review and update response and business continuity procedures at least yearly, and less than a third participate in a cyber wargame exercise. Given the incidence of cyberattacks, these types of preparedness exercises are key for adapting quickly and effectively to disruptions caused by cyberattacks.

How to Apply the Concepts from This Chapter: Conduct Your Own Wargaming Simulation

For the exercise in this chapter, we are going to simulate a ransomware attack. The intent of this exercise is to help you understand which areas of your security defense, processes and procedures, and resiliency you will need to bolster to effectively protect against one of the most severe scenarios you may experience. These questions are designed for you to run a simulation in your own organization, ideally with a cross-functional business and technology team, to determine the most appropriate responses for your organization and context.

While the first twenty-four hours of a security event are critical for the survival of your company, that is only the beginning. What should be clear, however, is that you must be able to act quickly to mitigate any damage, and you cannot act quickly enough if you have not prepared. While this example may seem dramatic, many companies described in this chapter have had similar events occur. This exercise should be a stark reminder of the importance of thinking about cybersecurity and your response before it happens to you. There are no clearly defined "right answers" of what to do and the order in which to take actions. This exercise is intended to help you start thinking about the breadth of issues you may need to consider in the wake of an actual attack. Let's begin.

It is 8:00 a.m. on a Monday.

"Hey, this is Brian from the Global Security Operations Center. We have a line of business security event in our health services business. As of 7:45 a.m., 10 percent of our team members are locked out of their computers. The numbers are growing rapidly, and we expect 50% impact within the next hour. The screenshot indicates that we have twenty-four hours to pay the $100 million ransom or our files will be deleted forever. I will send you a screenshot. We are enacting emergency protocol immediately and will have an active threat response bridge open 24/7."

You hop on the emergency conference line and learn the following about the current situation:

- The ransomware has already disabled 10 percent of your employee computers/systems and is projected to subsume 100 percent of your employee computers within the next two hours.

- Several mission critical applications are unresponsive and/ or inaccessible. The following areas of your business are impacted:
  - core business: severely degraded (you cannot safely serve customers)
  - finance: severely degraded (you cannot accept or process payments)
  - IT: moderately degraded (a few IT tools are inaccessible; however, security/monitoring remains unaffected)
  - other functions: TBD

- We do not believe that any data has been leaked yet. However, if the hackers have access to delete data, they may also be able to extract and distribute it.

What should you do next?

- Did you have a plan already in place to tackle the situation? Or do you need to create one as you go?

- Is there any data or other digital assets that must be prioritized?

- Should you divert customer traffic to functional sites while they are still available?

- When should you begin investigating the root cause, source, and remediation?

- How and when should you prepare communication to employees notifying them about the situation? Authorities? Media?

- How do you determine restoration, backup, failover timeline, and feasibility?

- Do you send noncritical employees home? Keep them on standby?

- What else do you need to answer to make an informed decision?

It is now noon . . .

- Interim workarounds to keep the business functioning are not working. There aren't enough functioning locations in the area/region to support spillover—moving over to competitors/ partner networks is the only option.

- A forensic investigation identified the source of the problem. A remote access trojan disguised as an attachment lay dormant until activated this morning. It managed to sneak past the existing antivirus software.

- Disaster recovery environments are compromised. Application restoration requires reimaged servers and reinstalled applications.

- Four weeks are required for full forensic analyses on all infrastructure.

- The call center is responding to major influx in contacts. They are messaging that the company is aware and investigating the issue and core business has not been compromised.

- The media has picked up this story.

What now? What is next for you and your company?
- Contact an external forensics investigator to accelerate investigations?

- Escalate with law enforcement?
- Identify and implement manual process alternatives? Add temps to perform tasks manually?
- Release media statement?
- Contact regulators?
- Hire a PR firm? Or scale up PR operations?
- Hire outside counsel?
- What else do you need to know?
- What choices must be made for next twenty-four hours? Weeks? Months? Years?

III   Organizational Building Blocks for Navigating Disruption

Work life balance gets thrown aside. Don't let it. Work life balance is not just a buzzy, self-help term that real business people laugh at. You need it. So, take care of yourself and enforce the boundaries that make balance possible, or at least try to. So, go to bed. For too long, we've believed that the most successful people are working 24/7. No. That's wrong. That's a recipe for burnout, and burnout hurts.
—Stephanie Ruhle

Not coming into the office every day sounds like a dream come true. No need to endure long commutes. No need to cram one's self into a tiny windowless cubicle where the noise from adjacent cubicles precludes the possibility of a confidential conversation or completing a task that requires intense concentration. No need to invest in a fancy wardrobe since no one sees what you are wearing (or at least not below the waist). At the same time that individuals are enjoying the benefits of remote work, organizations can "save on high-priced real estate and . . . hire applicants who live far from the office, deepening the talent pool."[1]

The concept of remote work is not new. More than thirty years ago, management guru Peter Drucker declared that "commuting to office work is obsolete."[2] Much of what fueled this prediction was the advent of new technologies that enabled workers to telecommute. But the real impetus for telecommuting was an acute disruption that dates back almost fifty years ago—the oil crisis of the early 1970s—coupled with environmental concerns and traffic congestion.

Vicky Gan of Bloomberg CityLab writes the following:

> The founding document of telecommuting was a 1973 book called *The Telecommunications-Transportation Tradeoff*. Lead author Jack Nilles, a former NASA engineer, proposed telecommuting as an "alternative to transportation"— and an innovative answer to traffic, sprawl, and scarcity of nonrenewable resources. . . . Research for the book began in 1973, in the midst of a national energy crisis. "Coincidentally, the OPEC oil embargo had begun and the object of our research seemed a little more pertinent nationally," Nilles told CityLab. Meanwhile, the Clean Air Act had just been passed in 1970. The term "gridlock" entered urban planning parlance, as headlines warned of an impending traffic apocalypse.[3]

Technologies that enable remote work continue to advance, ranging from cloud-based video conferencing to collaboration tools. However, the technology seems to be more of an enabler than a driver. In March 2020, acute disruption accelerated the move to remote or virtual work, which had progressed in fits and starts since Nilles and his coauthors first advocated for telecommuting. Along with this shift, the use of collaboration tools grew exponentially, and workers and managers learned that they could take advantage of technology to work in new and more effective ways that could boost their individual productivity and performance as well as their contributions to teams.

The acute COVID disruption has let the genie out of the bottle with respect to new ways of working, and we see work changing for the future no matter how the pandemic plays out. As with the acceleration of digital transformation, organizations have an opportunity to seize this moment and intentionally design and shape the way their employees work—to optimize efficiency and productivity, yes, but also to be more data driven, to increase innovation, to foster connections, and to bolster well-being. They have a chance to reinvent work not just for individuals but also for teams. And while most people think of knowledge and office work as being primary candidates for reinvention, the pandemic-induced rise in telemedicine demonstrates that there are opportunities to rethink other types of work as well, including how we interact with customers, as we will see in chapter 11.

The Remote Work Revolution

One byproduct of remote work is that it provides a considerable amount of data that allows us to examine how it changes working patterns. For example, research suggests that the shift to remote work has lengthened the workday. Evan DeFilippis and his colleagues studied the electronic communication and scheduling records of over three million workers, finding workers attended approximately 13 percent more meetings and that those meetings had about 13 percent more people in the eight-week period following some form of lockdown in sixteen cities worldwide. Meetings were approximately 20 percent shorter, however, meaning that people actually spent *less* time in meetings, even though they attended more of them.[4] We expect that one explanation for this outcome was the relatively lower setup cost for meetings. When people do not need to be physically present to attend a meeting, it becomes easier to have more of them and each meeting doesn't need to last as long. Another explanation could be that instead of being able to walk down to the hall to talk to a coworker briefly about an issue, workers now need to formally schedule that time—and for the first time have digital traces of those interactions.

Microsoft used similar data to examine the behavior of 350 of its managers in China during the early lockdowns. They found that managers spent nearly twice as much time on virtual meetings and sent nearly double the number of instant messages to keep up with and motivate employees. They found that employees were working longer hours, with 50 percent greater instant messaging activity between 6:00 p.m. and midnight. Employees also tended to work more on weekends, with employees who previously logged few weekend hours working nearly triple the amount of time.

We suspect, however, that these longer hours working was also offset by people taking increasing amounts of personal time during the day with families all locked down together, a practice sometimes termed "time-shifting."[5] Ben Waber of Humanyze supports this perspective, saying "It doesn't appear that the amount of time people are working is changed, but it does appear like it's stretched out, which is not shock-ing." Not only were employees working from home but so were their spouses and children, creating different kinds of productivity chal-lenges than even seasoned remote workers had previously experienced.

Krishna Cheriath of Zoetis describes one such situation.

> One of my colleagues is a young mother with two kids in elementary school, who she has to help with online school from 10 to 11:30. How do we give her the flexibility and empowerment to design a custom work around those constraints, and then communicate them to her colleagues? Those are the kind of custom solutions where we need to help managers get better. It's about enabling empowerment, so that employees can design their own work styles. This is the new norm.

Yet, exactly what that future workspace looks like is not entirely clear as of this writing, and we expect it will not be for several years as com-panies struggle to find the right balance between colocated and remote work. Although a survey by Cushman & Wakefield finds that approxi-mately 75 percent of respondents indicated that they are being produc-tive working from home,[6] and companies like Facebook and Twitter indicate that certain employees may never need to return to the office,[7] many of the productivity advantages of remote work may not be sus-tainable for extended periods. The key is to take lessons learned from the 2020 worldwide experiment in remote work and use it to continue to evolve and intentionally design the environment and tools that enable your organization's talent to deliver the best outcomes.

It is important to learn about the pros and cons of remote work, because—based on our conversations with numerous business leaders and chief strategy officers—a significant portion of the workforce will not be going back to colocated work full-time for the foreseeable future, if ever.

The Benefits of Remote Work

Perhaps the biggest benefit of remote work that companies learned from the acute disruption outbreak is simply that remote work is doable. Frankly, we find it amazing that so many legacy companies were able to pivot successfully to remote work in such a short period of time. It may not have been easy or pretty or fun, but it *happened* and happened successfully. André Heinz, former head of global human resources at Siemens Healthineers, notes this success: "The most unexpected thing that has come up over the crisis is the resilience and flexibility of the organization and our ability to work full blown without disruption, despite everyone being on lockdown." Monty Hamilton of Rural Sourcing notes that "feedback from our customers is that our productivity hasn't dropped. If anything, it's gone up a bit." David Quigley of Boston College also shared his surprise at the effectiveness of remote work: "I know individuals have the ability to adapt, but the ability to successfully do that collectively is another matter entirely."

This successful transition to remote work generally supports our thesis that people are the real key to digital transformation. When digital transformation became a necessity during the acute COVID-19 disruption, people were able to adapt and the technology was there to support them. Furthermore, people have been generally pleased with the shift to remote work, with nearly 83 percent of workers reporting that they wanted to work from home at least one day per week and over half of managers reporting that they are supportive of some work from home once a return to colocated work is possible.[8] Kristin Darby of Envision Healthcare reflected on the medical group's plans to return to the workplace. She says that "we have a return to work committee that is continuously evaluating the risks and putting the health and safety of employees and the public first.

It is important to note, however, that the reaction to remote work varies across age and in ways that may be counterintuitive. Although we would expect that younger, more tech-savvy people would be better able to pivot to remote work, the opposite is actually true. Research by the architectural firm Gensler finds that older workers are generally more satisfied with remote work, primarily because they often have bigger homes that are more conducive to remote work and have older families at home. Younger workers, by contrast, are less likely to have favorable conditions for remote work and are more likely to be living in smaller apartments in urban settings, either alone or with roommates they may not know well or with small children whom they can no longer take to daycare or school. Our interviews reveal anecdotal evidence for this trend. Silke Sasano, principal key expert design thinking at Siemens Healthineers, observes that "some members of my team were very young, living on their own in single apartments. Sometimes they were very lonely. You could feel this sadness being so isolated, so we tried to come up with social outlets."

Furthermore, younger workers are also more likely to be at earlier stages in their careers and may find it harder to build their networks or to navigate and learn a new organization without walking the halls. It will be important to monitor employee satisfaction with remote work over time, because enthusiasm for remote work may wane the longer it persists.

Not only is short-term productivity stable with the shift to remote work, but for many workers productivity actually goes up. When employees do not need to manage a long commute, they can dedicate some of that commute time to productive work. Forty-one percent of respondents say they are actually more productive during remote work than they had been before, and 31 percent say they are equally as productive.[9] Christine Halberstadt of Freddie Mac echoed this sentiment: "We've been very pleasantly surprised at how people have been able to maintain productivity."

Of course, we should caution that more research is necessary to determine what part of that productivity gain was due to remote work and what part was fueled by the pandemic. When employees in lockdown have few options for outside activities or entertainment, combined with a fear of losing one's job due to a recession, it may be that they simply work harder. We are more comfortable concluding that—under the right conditions—remote work can be at least equally or as productive as colocated work. Indeed a 2015 Stanford research study of a Chinese travel agency found a 13 percent increase in productivity with remote work.[10]

Certain types of employees are more likely to thrive than other types of employees. Some have suggested that employees with introverted personalities are more likely to thrive in remote work settings than employees with extroverted ones.[11] If so, this could result in a shift in workplace dynamics, since research indicates that extroverts have an advantage in colocated work. Furthermore, organizations' embrace of remote work may open up new work opportunities for people on the autism spectrum,[12] which the Centers for Disease Control estimates as about 2 percent of the US population. Employees with greater openness—a basic personality trait denoting receptivity to new ideas and new experiences (often associated with a growth mindset) are also more likely to thrive during remote work, possibly because of their general willingness—to experiment with new ways of working, as are people who like to make quick decisions compared to those individuals who like long conversations.[13]

Companies with a significant percentage of remote workers pre-COVID who previously felt excluded from the office say they no longer feel that way. Eric Schuetzler of Beam Suntory describes this phenomenon.

> In the past, if you're outside of the main location, you'd feel lesser in many different ways. Now, everybody has an idea of what it's like to not be right next to everybody else. It's not about Zoom or Microsoft Teams, it's about the actions and behaviors you bring to the meetings. You need to have an understanding of the experience that somebody will have on the other side. The result is, while no one's actually together, people actually feel more together. It's a sense of virtual inclusion.

The distributed nature of remote work is described as a great equalizer, putting everyone on an equal playing field, which can be promising for diversity, equity, and inclusion efforts. And with the virtual nature of meetings, it is easier to attend more meetings, as location, travel, and cost are no longer limitations. Some respondents indicate that remote work leads to a less hierarchical organizational structure. André Heinz continues: "In this crisis, our decision-making processes have incredibly accelerated. It's also contributed to a flatter hierarchy because everyone is working from home. There is no such thing as a corner office or a company car now. There's a lot more cross organization collaboration."

The Drawbacks of Remote Work

Of course, remote work may not be entirely positive. There are elements of face-to-face meetings that cannot be replicated online. One drawback is that remote work may decrease innovation. Academic research on knowledge management dating back decades notes that new knowledge is typically created through the combination and exchange of existing knowledge.[14] Much of this valuable combination and exchange of knowledge in today's organizations tends to occur in serendipitous interactions when employees are simply in the same place, connections that could not happen without colocated work. Brad Keller of Humana notes this implication of remote work: "When you don't have physical presence, you miss casual collisions that are really crucial. If you're sitting in a workspace and you hear someone talking about an interesting project, you may realize that you're the right person for that project. You can't replicate that experience over video."

Other research on social networks points out that valuable knowledge is more likely to be found through casual or "weak" connections than strong relationships,[15] since workers with strong connections already share much common knowledge. The workplace analytics firm Humanyze indicates that its data shows the rapid deterioration of these types of "weak" ties that have long been shown to be the most valuable source of new information.

Humanyze's Ben Waber describes how this shift in relationships as work shifted to remote shows up in their data. Their data showed that connections with people you already worked closely with actually got stronger. He explains that "communication with close collaborators, people that you spend an hour or more with a week, has actually gone up. Many more people that fall into that category now, too. People who were on the borderline—that you met previously 15 minutes to 45 minutes every week—have been elevated to close collaborators."

Paired with this strengthening communication between close collaborators was a significant drop off in communication with other people. Waber continues: "For relationships that fell below that line, those communications have dropped precipitously. This drop off represents one of the challenges for companies over the next couple years. The people that you used to bump into the hallway or were acquaintances at work, you are communicating with them a lot less." Since these serendipitous connections are often the source of new ideas, the impact of this change in communication could have significant implications for an organization's ability to innovate. Waber concludes that

> this decrease doesn't impact work right now, but it will in the longer term. Decades of research shows that these relationships provide a strategic understanding of what different parts of the company are doing, help generate new product or service ideas, about really peripheral dependencies, which is where bugs pop up and real product misses come out. This shift is not about near term performance, it's going to be about longer term.

Another disadvantage to remote work may be the deterioration of company culture. Culture is a critically important factor driving digital transformation and innovation. Our studies show that digitally maturing companies are more likely to have a culture characterized by agility or nimbleness, risk taking, experimentation, collaboration, and cross-functional teaming. They are also more likely to report investing time, energy, and money in developing these characteristics and report using these features as the primary way they drive innovation. Yet, a survey by Cushman & Wakefield finds that 50 percent of remote workers have trouble connecting to company culture during remote work. This survey also finds that human connection and social bonding are suffering, impacting connection to corporate culture and learning. The survey's finding is shared by André Heinz, formerly of Siemens Healthineers.

> Identification with a company happens through personal interaction. My concern is if we do more remote work, this identification may disappear. How do we nurture this identity while separated from each other? How do we make sure that this mortar between the bricks that binds us together is still nurtured? I think if we don't consciously think about how to make that happen, we will create a risk and only realize it too late.

Monty Hamilton of Rural Sourcing noted the impact on culture more immediately. He said, "We have a really strong culture. We put 125 creative people in an office. Now, we're suddenly working from home for an extended period of time. We're doing everything we possibly can to continue to create and maintain our culture and our community."

Humana took some immediate steps to attempt to overcome the impact of remote work on culture and connections, building an outdoor office space for employees to use. Keller's team secured use of a large park near their Louisville, Kentucky, headquarters and set up tents, power, office furniture, Wi-Fi, and bathrooms in a socially distanced way—even scheduling food trucks to be on premises. They then set up a reservation system that allowed employees to schedule use of the space and control capacity. The reservation system also provided employees the opportunity to see who else would be at the workspace during that time, allowing them to be more intentional about making use of the time to connect with colleagues.

Michael Aldridge of Humana noted this benefit, saying, "Looking at the list, I realized that there were about 5–6 people who would be there at the same time as me that I hadn't seen since the lockdown began. It allowed me to just have 5-minute conversations and rekindle some of those connections." Ninety-five percent of Humana's employees indicated that the outdoor office spaced helped support their work, with the ability to meet colleagues at a safe location and change of routine being the biggest benefits.

Finally, remote work may not be conducive to mental health. Forty percent of respondents say that their mental health has deteriorated during the lockdown.[16] Another survey finds that only 54 percent of respondents have a feeling of well-being during extended remote work. Janet Pogue McLaurin of Gensler shares that "a recent Gensler survey found that 74% of respondents said that they missed people the most. They missed meeting with their colleagues and the impromptu connections." Additionally, remote work may further blur the boundaries between work and home, making it easier to always be on and harder to get the downtime needed for well-being.

Just as in our previous section, we should be careful to note that it is difficult to tease out which parts of these drawbacks should be attributed to remote work and what part should be attributed to the experience of the acute COVID disruption in general. It may be that people handle remote work perfectly well if they can engage in normal human relationships outside of work and are not experiencing isolation or the fear of sickness or economic calamity. Alternatively, it may simply be that being in virtual meetings all day does have a fundamental effect on our mental well-being, as implied by the number of articles providing tips for combating video-conferencing fatigue. More research, as well as additional experimentation, may reveal the answers.

What Will Offices Look Like?

If the future of work is likely to involve both colocated and virtual processes, what does that mean for the future of the office? Clearly, if some degree of shift to remote work persists, it will change the commercial real estate picture considerably. Brad Keller of Humana says, "I think there will only be a few people who would like to go back to the in-office-all-the-time model. People won't need a place to go five days a week, but rather a place to go to get work done. In the long run, I think companies will need less real estate, but we were already realizing that before the pandemic."

During the acute COVID disruption, the retailer REI sold its new eight-acre headquarters before ever occupying it: Peter Grant reports in the *Wall Street Journal*, "The outdoor clothing and gear retailer . . . had been planning to occupy the 400,000 square foot campus in Bellevue, Wash. . . . [in the summer of 2020]. REI designed the elaborate complex to reflect its outdoorsy image, and the company once hoped it would serve as a way to recruit new employees. The property features outdoor staircases and bridges, a courtyard of native plants, and skylights to let in sunshine and air."[17]

REI concluded that the new space was no longer consistent with how people would work in the future: "The company's work-from-home program also proved to be surprisingly successful, said Ben Steele, REI's chief customer officer. REI will rely more heavily on remote work and smaller satellite offices."[18] Facebook, however, bought the REI property, which is next to three other Facebook buildings under development. Even as Facebook acknowledges that many of their employees will shift to working from home, their appetite for real estate has not diminished entirely during the pandemic.

Cushman & Wakefield believe that—at least for the short term—overall square footage will likely remain the same, balancing social distancing's relaxing of space density with less office space headcount demand in the new total workplace strategy.[19] We think it is likely too early to determine the long-term effects on corporate real estate, because it will involve a complex set of economic conditions that are simply too difficult to predict at this point.

Nonetheless, this large-scale remote work migration gives companies an opportunity to redesign their physical workplaces with more intentionality, based on recent shifts in behavior and continuing advancements in technology. Janet Pogue McLaurin of Gensler says, "Over time, workplace effectiveness was starting to decline—it actually declined a lot from 2019 to 2020. This crisis may be an amazing opportunity to step back and think about how we fix what wasn't working before: rethink the workplace, rethink office buildings, rethink that whole experience. Now we're in the process of reimagining."

Cushman & Wakefield suggest that office space will be reimagined into a "total workplace ecosystem," arguing that offices will no longer be a single location but instead multiple locations and experiences to support convenience, functionality, and well-being. These offices will be designed to optimally support the type of work that people will be doing in-person. The purpose of the office will be multifaceted, focusing on the type of interactions that in-person work will be conducive to—strengthening organizational culture, learning, brainstorming, onboarding new joiners, strengthening relationships among employees and with customers and colleagues, and supporting innovation.

André Heinz echoes this sentiment, saying, "We're now thinking of a completely different office structure with more collaborative spaces. We don't need as many single offices anymore, since many people are likely to work at home." At the same time, companies may need to retain a certain amount of space for private individual work, as some workers simply do not have the space or privacy to work from home.

Sara Armbruster of Steelcase, which produces architecture, furniture, and technology products and services for offices, describes the value of designing a flexible work environment: "We introduced last year a collection of flexible furniture elements where components are on casters and can be moved around and arranged by workers to adjust for their needs on any given day." A flexible work environment allows teams to configure the space to meet their needs, whether it is spacing further apart to maintain social distancing or pushing the components further together for close group work. The key is to imagine the new ways of working and to design a physical space that best serves your organization's evolving needs.

Rethinking Work through Continual Learning

The shift of many jobs to remote settings requires many workers to fundamentally rethink the critical components of their job and where they add value. Interestingly, many of the jobs that are least susceptible to the chronic digital disruption are greatly impacted by the 2020 acute disruption. For example, in the early 2000s the revenue model of the music industry shifted from a focus on recorded to live music to respond to the digital disruption ushered in by Napster and other online music sites, a shift that has made it more vulnerable to this acute disruption. The job that McAfee and Brynjolfsson note as one of the least susceptible to digital disruption—massage therapist—is among the more heavily affected jobs during the acute COVID disruption. While the public narrative is that the disruption is split along class lines, with white-collar work being less disrupted than blue-collar work, the reality is far more complex. Many high-paid white-collar workers, such as doctors and dentists, depend on in-person, high-touch services, which have been significantly disrupted.

We believe that many employees will need to rethink their jobs in the light of the shift to remote work. One way that employees can do that is through continual learning and adopting a growth mindset. Not surprisingly, demand for and participation in online skill development has increased significantly during the pandemic. Mark Onisk of Skillsoft notes the rise of digital learning that occurred during acute disruption.

> There are four key elements why we feel digital learning is just a better way to learn. Firstly, digital learning has a tendency to be far more purposeful, giving you the ability to filter and recommend skills-based paths more than you ordinarily would with in-person training. Secondly, you're able to get to content that is far more engaging and immersive, faster, because it's self-driven. Thirdly, you can offer more hands-on applications with digital learning so people can practice skills in a meaningful way. Finally, it's far more scalable, with advances in AI, machine learning, and other technologies, personalized learning journeys can be defined at a magnitude that traditional methods are unable to match.

But having a growth mindset is more than just attending online courses and formal institutional learning, which often takes place in classrooms and training programs. It inspires and requires a different style of learning, one that is necessarily more self-driven and less structured. Institutional learning and formalized training alone are unable to keep up with the pace of change and technological development. A different type of learning is needed: one that is continual, experiential, and exploratory. It is about finding and seeking learning in nearly everything—being in a constant state of growth.

To foster this mindset in their organizations, leaders must build a culture of continual learning that defines learning broadly. In addition, they must recognize, encourage, and provide opportunities and access to all of the different ways people learn. This includes formal programs and courses as well as the many informal ways to learn, such as attending conferences and webcasts, listening to podcasts and TED talks, and reading articles, reports, and books. Having formal and informal coaches and mentors is also a critical way many people learn. In addition to this, people learn by doing, so it is important to create opportunities for employees to learn on the job through new roles, projects, challenges, and exposure to different people, problems, and perspectives. Finally,

Table 9.1
Going beyond the classroom for continual learning (examples)

Institutional	Informal	Experiential	Exploratory
• Degree and certificate programs • In-person training courses • Online learning	• Conferences • TED talks • Podcasts • Webcasts • Articles/research reports • Books • Coaches/mentors	• Opportunities for new roles, projects, and challenges • Exposure to different problems, people, and cultures • Engaging in external collaborations	• Engaging in experiments, big and small • Creating new knowledge

people can engage in experiments big and small to discover new knowledge. This is a key mindset and muscle that organizations can build to help them get better at testing, learning, and experimenting. See table 9.1 for examples of the different ways people learn.

With respect to our research on chronic disruption, we found that the key differentiator between early, developing, and maturing companies is not necessarily that these companies have better talent but that they are doing more to help their employees be better equipped for working in a digital world. We expect the same will apply to acute disruption: the companies that emerge stronger are those that help their employees gain the skills to work in the "next normal" and beyond.

Shaping Ways of Working into the Future

Dozens of articles have recently been published on remote work, with a focus on how to lead remotely. By the time this book is published, many more will have been written and many readers will have considerably more experience working remotely than they do as we write this chapter. We have no certainty as to what the remote work situation will look like when this book hits the shelves. Will we have returned to some semblance of colocated work, or will remote work continue to be the norm? What advice can we give that would not be obsolete or dated given the number of possible futures we face?

Our only advice is to allow your ways of working to continue to evolve through and after the disruption. People have spent decades or more learning to work colocated, and whatever processes your organization developed in March and April of 2020 are almost certainly not the best way to work virtually. Many organizations have simply lifted and shifted the physical ways of working into a virtual setting, simply replacing in-person meetings with video conferences. Having proven that they can do many things virtually, companies are presented with an opportunity to build on what they have started and to shape the future of how they work. We noted early that some of the pros and cons of remote work are not yet clear, and we fully expect that list will grow and shift in the years to come. Nevertheless, we are confident that colocated work will be superior for some types of tasks and remote work will be superior for others. The trick, of course, will be knowing which is which.

A great example of the evolution of virtual work comes from the marketing services conglomerate, WPP. Jacqui Canney describes trying to figure out what to do with an existing three-hundred-person intern program that could no longer be hosted in the traditional face-to-face setting.

> We brainstormed an idea for a 10-week curriculum that had a different theme every week. We invited the 300 people who we had already tapped as interns, but we decided to open it up to anyone. What we thought was going to be 300 turned out to involve over 850. And it was 55% people of color, 71% women, and over 300 universities from about 30 countries around the world. The feedback was overwhelmingly positive. Now we have a whole way of connecting to these almost 900 people that we wouldn't have had a relationship with before. The whole thing just evolved into something very different than when we started and opened up lots of opportunities. Now we're thinking how we even up our game again. I think it's changed the way we approach this forever; I really do.

WPP is now thinking about how to evolve this program further, potentially extending it to a year-round effort, providing academic credit, and opening it up to partners. Yet, it's safe to say that it would never have started down this path if it were not for acute disruption. When the barriers of colocation are removed, it often provides new ways to approach old problems in sometimes superior ways.

Acute disruption presents opportunity for innovation in work, and we hope organizations will continue to take advantage of it. It may even be helpful to shift the thinking from remote/virtual work versus colocated work to a concept that blends and creates a new mix of physical and digital workplaces, where the digital workplace becomes more than just a way for people to work from different locations. There is much opportunity to improve the digital workplace from the current typical collection of assorted desktop tools that aren't well connected. With advancing technologies, creativity, and will, companies can rearchitect an integrated digital and physical workplace where people can work, collaborate, and connect in new ways. The key is to use the lessons from 2020 to intentionally design the way your talent works, lest you find yourself still operating the same way in another ten years.

We started this chapter with an excerpt from a commencement address delivered by NBC News senior business correspondent, Stephanie Ruhle, in 2017 at her alma mater, Lehigh University. In her commencement address, Ruhle talks about the importance of avoiding burnout. As we continue to figure out how to work in the "new normal" and as the enabling technologies continue to evolve and improve, companies have an opportunity to rethink the way their employees work to take advantage of technology and optimize human potential, including productivity, connection, well-being, and avoiding burnout.

How to Apply the Concepts from This Chapter: Starter Questions for Work, the Workforce, and the Workplace

We recommend that leaders consider three elements of the future of work when thinking about how to adapt to disruption:

1. **Work** (what can be automated): How to use AI and robotics to complement human-delivered outcomes
2. **Workforce** (who can do the work): How to find, leverage, engage, and develop new forms of talent
3. **Workplace** (where is the work done): How to manage collaboration, teaming, digital technologies, and the work

Adjusting to new forms of work will require rethinking talent models, teaming (a topic we discuss in the next chapter), and transitioning away from traditional organization structures. The challenges around workforce, workplace, and work are big ones, but let's approach them in stages.

Rearchitecting Work

With advances in AI, robotics, and other technologies, there is an opportunity to rearchitect the way work is done, shifting certain types of work performed by humans to machines. By doing so, organizations can use technologies and elevate what humans do (see "Age of With" in chapter 7), applying digital tools to the work itself and focusing on outcomes. Here are some starter questions:

- What are the work outcomes desired? How can they be enhanced by technology?
- Which technologies are most likely to impact those outcomes in our company or industry?
- Which organizational roles are most likely to be affected by emerging technologies (e.g., through automation or augmentation)?
- What will be the impact of these new technologies on roles, resources, or value (e.g., reduction of headcount relative to workload, augmentation/enhancement of roles, training requirements)?
- Where are the highest impact opportunities?
- What exogenous factors (e.g., labor laws) could potentially change our responses to these questions?
- How do we manage concerns among our current employee population?

Unleashing the Workforce

Our previous studies show a need for talent and skills within most organizations. Creating a culture of continuous learning and developing existing talent are critical parts of addressing that need. However, organizations must also think about expanding their talent models from

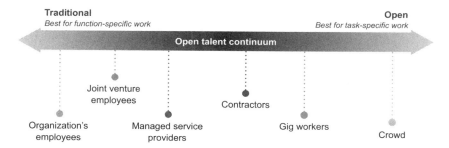

Figure 9.1

The talent market covers a spectrum of worker types and work arrangements.
Source: https://www2.deloitte.com/us/en/insights/focus/technology-and-the-future
-of-work/redefining-work-workforces-workplaces.html

traditional full-time employees, especially for hard-to-hire skills (e.g.,
software developers) or where demand for work isn't clear yet (e.g., block-
chain). How can companies unleash their workforce—both by unleashing their full potential and by untethering them from traditional models
and physical-only workplaces? This requires connecting to talent differently (digital tools, new ways of leading) and tying them to your organization and its mission through engagement, contribution, impact,
and purpose (see chapter 3 on purpose, values, and mission). Figure 9.1
shows various types of workers in the emerging workforce.

In thinking through the right mix of traditional and open talent
roles best suited to your needs, reflect on some key issues. Here are
some starter questions for unleashing the workforce:

- How do we create a culture of continuous learning and continually
  upskill our workforce?
- What are the skills that we'll need in the near and far future?
- What are the roles or skills that may not be accessible through our
  traditional approach of hiring full-time employees?
- What other talent models should we consider? What are the pros
  and cons associated with each model relative to the potential gaps
  we are anticipating?

- How are we positioned as an organization to implement these alternative models? Where do we need to build new capabilities and skills?
- What exogenous factors (e.g., immigration/visa policies, labor laws) could potentially change our responses to these questions?
- How do we manage concerns among our current employee population?

Adapting the Workplace

The workplace—whether physical, digital, or a combination of both—is critical to building relationships at work, finding meaning in the work one does, and is highly correlated with innovation.[20] We have already discussed how virtual working is likely here to stay for some types of work and some organizations. For others, however, the nature of some work restricts physical proximity (e.g., hardware engineering), and still others are being fundamentally redesigned. While there is a real opportunity to be both cost-effective and innovative, leaders must be thoughtful and deliberate about determining what work is best suited to physical interactions, what work is productive when performed virtually, and where a balance is needed. Furthermore, they need to think through how to better integrate the physical and digital and use technologies to enhance both environments. Figure 9.2 shows at a high level how workplaces are shifting from collocated to more distributed formats.

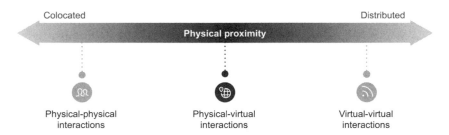

Figure 9.2
Adapting the workplace to a shifting environment.
Source: https://www2.deloitte.com/us/en/insights/focus/technology-and-the-future
-of-work/redefining-work-workforces-workplaces.html

Below are some starter questions to explore the possible mix of collocated and distributed work:

- What work requires physical interactions and to what degree? What work can continue to be performed virtually? How does this impact various roles, which may consist of work that can be performed across digital and physical environments?
- How do we use technologies to create better digital and physical workplaces?
- What exogenous factors (e.g., transit infrastructure) could potentially change our responses to these questions?
- How do we manage concerns among our current employee population?

10 Teaming Your Way through Disruption

Alone we can do so little; together we can do so much.
—Helen Keller

We know from our prior research into chronic digital disruption that teams are incredibly important when it comes to an organization's digital maturity and its ability to adapt to a digital environment. We also know that the most effective teams are cross-functional. The most advanced group, the digitally maturing companies, depend on cross-functional teams to accelerate their innovation efforts. Eighty-three percent of digitally maturing companies say they use cross-functional teams, compared with 71 percent of developing companies on the digital maturity spectrum and 55 percent of early-stage organizations (see figure 10.1). We find that the increasing reliance on cross-functional teams is associated with a decrease in the perception that organizational processes interfere with the organization's ability to be nimble.

A cross-functional team starts with people from multiple departments. Rather than answering to whichever line manager they're officially assigned to, the team might answer to a project manager or a corporate innovation executive. In addition to the adaptability noted above, survey respondents note that a key advantage to cross-functional teams is the enhanced access to resources, such as diverse perspectives, broader skill sets, and new ideas.

Digitally maturing organizations operate cross-functional teams differently than early-stage companies. *(% of respondents who agree or strongly agree)*

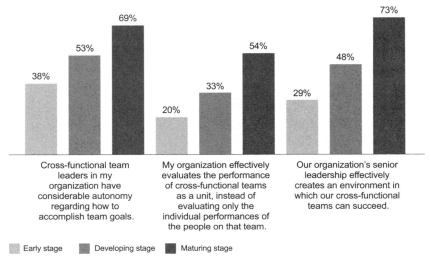

Figure 10.1
Hallmarks of empowered cross-functional teams

The differences among maturity groups are even more pronounced when respondents are asked *how* they deploy cross-functional teams. Executives and managers at digitally maturing companies, compared with developing and early-stage ones, say these teams are more likely to have considerable autonomy regarding how to accomplish goals (69 percent versus 53 percent and 38 percent, respectively). Shamim Mohammad of CarMax underscores this approach to managing cross-functional teams. He says that "one of the benefits of the empowered teams is that we set the strategy, but we empower them figure out how to get it done. Our teams experiment directly with customers and associates to test and learn, which enables us to build and deliver the capabilities they truly want and love."

Another important aspect of cross-functional teaming is that senior leaders create a supportive environment for their teams and evaluate them as a group. Our survey data shows that 73 percent of digitally maturing companies create an environment where cross-functional teams can succeed, compared to only 48 percent of developing companies and 29 percent of early-stage companies. Mohammad describes this supportive environment, saying that "a key reason our teams are thriving is because they know that the company has their backs and the company is providing them the support they need."

These teams drive innovation in similar ways in both chronic and acute disruption by increasing the ability of the organization to sense changes in the environment and respond quickly to them. Mohammad articulates the benefits of empowered teams for dealing with acute disruption: "It allowed us to move really fast during this pandemic. Our teams know the mission, and they have the flexibility and ability to change and adapt quickly to be there for the customers and associates."

Barriers to Cross-Functional Teaming

Operating via cross-functional teams may pose new kinds of management challenges, however. More than half of our survey respondents cite problems with team alignment and an unsupportive culture as the biggest barriers facing cross-functional teams. To overcome these challenges, companies must secure buy-in via clear and frequent communication with employees, says Matt Schuyler of Hilton.

> We see great benefit to our business and our culture when our team members work cross-functionally, so we spend a lot of time challenging our teams to think outside of their own subject-matter expertise. To encourage a more collaborative mindset, we invest time reminding them that you can have it all here and that communicating outside your silo is in some ways more important than within your silo.

Michael Arena, former chief talent officer at General Motors, cautions that while cross-functional teams are an important source of innovation at his former company and other digitally maturing ones, they are not a panacea. Innovation is a process, and it occurs in stages. "For organizations to be adaptive," he says, "the very first thing we need to do, especially as we're talking about org design and practices, is to ditch the one-size-fits-all mindset."

Arena studies organizational network analysis and the impact of organizational design on innovation. He notes that cross-functional teams may be brought together to address one aspect of innovation (say, ideation), but team members may have a different role when it comes to other aspects of the innovation process (say, diffusion). "It could be that for six weeks we're pulling people together for a specific purpose," Arena explains. "They've got these milestones and for six weeks they're dedicated to getting something across the finish line. And that's the design for that six-week interval. Then those team members are going back to their steady-state jobs where we're going to ask them to help diffuse this out across the broader organization."

Daniel Leibholz of Analog Devices describes a similar way of working with distributed teams at his company. "Our design and product teams have a lot of freedom to explore their unique spaces, but in a fairly distributed manner, so that they stay close to their customers and hone their performance edge. We connect these teams through technology platforms, shared methodology and a strong culture of collaboration to create an organization that can learn and adapt very quickly."

Collaborating across Boundaries

Unique to the acute COVID disruption, however, is that many of the barriers that were broken down were between organizations, not just within the organizations themselves. Many companies recognized that the COVID threat was a crisis that superseded competitive concerns and began to pool resources and insight in unprecedented ways.

One industry in which this cross-organizational collaboration occurred was the healthcare industry, where companies began working together to address the shared global health crisis. Krishna Cheriath of Zoetis (formerly of Bristol Myers Squibb) explains that

> when I was at BMS, I found that our competitors became partners and collaborators. There was a coordinated effort across biopharma to look at our respective assets list to see which could be potential candidates for vaccines and cures. We asked how we can pull together scientific talent to progress the

promising targets, irrespective of which companies they belonged to. When they do show promise, how do we scale manufacturing to the levels that are needed to bring this at scale globally? It is going to require the capacity of not just one company, but unprecedented levels of biopharma collaboration.

A similar story played out in the spirits industry, as companies sought to step up to address the shortage of hand sanitizer in the early days of the pandemic. Beam Suntory knew how to make spirits but not hand sanitizer. It had to figure out how to do so in a way that was safe and met changing Food and Drug Administration and Alcohol and Tobacco Tax and Trade Bureau (TBB) regulations. Though it had to make many logistical shifts, the company was able to leverage a huge cross-functional team, set up the proper systems for governmental and regulatory documentation, and convert an existing pilot plant facility to begin producing the sanitizer in two weeks.

To get up to speed quickly, the company worked directly with its competitors to figure out together how to accomplish this shared goal. Eric Schuetzler, global head of research and design (R&D), describes the experience: "We don't often find ourselves on phone calls with our top competitors and their R&D leaders—but we did. We were on multiple times a week, sharing information and insights. Normally, we would never do that." The competitors worked together to figure out how to pivot manufacturing to develop the needed product, such as sharing where to source needed raw materials.

Regulation also presented an issue because unless the alcohol used to make hand sanitizer was denatured, they would need to pay federal excise tax, even if the sanitizer was donated. As Blake Layfield, senior manager of global R&D, explains, "To ensure compliance, we created denatured alcohol by adding the most bitter compound on earth to it, ensuring it could not be consumed." Schuetzler concludes his experience, saying, "It was very unique, and a very different way of working. It was very seamless and collaborative, and all about the greater good for everybody." They were only expecting to produce the sanitizer for four weeks but ended up producing it for thirteen, resulting in 41,000 US gallons. The product would have been valued at $2 million given market prices, but Beam Suntory donated the vast majority, reserving a portion to protect the company's employees in operations and offices.

Others reported working across boundaries to collaborate with leaders in similar situations. For example, David Quigley of Boston College notes the various external networks that he used to attempt to prepare the organization for a return to in-person learning.

> A number of external networks—whether for other provosts or just other local universities—were really helpful for our planning and for helping keep my sanity. It was striking that there's no one right solution—so much depends on which peer set you're looking at for what might become a preferred solution for one of those groups. It was helpful to hear how other folks were approaching similar challenges, because there is no playbook and nobody's done this for 100 years. We're all figuring it out and trying to come up with the right solution for our particular community and our institutional setting.

Why Rely on Cross-Functional Teams in Response to Disruption?

As we noted in chapter 1, the reliance on cross-functional teams is a surprising connection between chronic digital disruption and the acute COVID disruption. In understanding why cross-functional teams are essential to responding to acute disruption, we found an article by Elaine Pulakos and Robert B. Kaiser to be helpful in framing the different ways that teams can work together[1]. They note that an important factor in determining how teams should work together is the degree of novelty in a problem or situation. They identify four types of teamwork:

- When problems are routine and well understood, the simplest form of teamwork is simply *handing things off.*
- More complex and less routine problems may require *synchronized work*, where team members perform tasks independently but communicate and coordinate on progress.
- The next level is *coordinated work*, where a team works together but each member has clearly defined roles.
- Last, truly complex and novel problems may require *interdependent work*, where a group of people come together simultaneously to address a novel problem or situation.

We expect this framework helps explain why cross-functional teaming is associated with many companies' responses to acute disruption. When companies are making decisions in response to a truly novel situation, they need decision makers from various areas present to bring all areas of expertise to bear. The company needs to be nimble, and small working groups allow them to do that. The acute COVID disruption also temporarily removes some of the physical and cultural barriers that previously may have gotten in the way. Geographic barriers are no longer an issue, and employees who may have worked in different offices or cities can all gather remotely and work the problem together.

Florian Hager, principal key expert at Siemens Healthineers, notes another reason that cross-functional teams have been effective in responding to acute disruption. He says that "when you have a new team, you often have a team building event over a day, a weekend, or even a week. People just play the game because they know after that event, they go back into comfort mode again. The interesting thing is this situation has become a 90 day or 3–4 month team building event."

Pulakos and Kaiser encourage teams to continually evaluate the types of work they are doing in terms of these four categories. They then continually refine the work processes to become simpler as the problem becomes more well-defined and work can be decomposed into distinct parts. Numerous instances of interdependent work are necessary to help navigate novel situations, and teams can then simplify to more independent ways of working as they figure out the environment.

Learning Cheap and Fast

Another potential benefit to cross-functional teaming in response to disruption is the opportunity it provides to be more experimental. As companies face a novel business environment, they need opportunities to try new things to determine what works in that environment. Cross-functional teams allow the company to test and learn in the new environment to help find ways to respond.

Prior to the acute COVID disruption, CarMax—the Richmond, Virginia, auto retailer—had embraced cross-functional teams so thoroughly that its technology organization has basically dispensed with traditional planning. Instead, Shamim Mohammad says that CarMax expects innovations to bubble up through its product teams. "If you think about how fast technology is changing and how fast customer expectations are changing, to deliver what the customers are looking for, you have to organize as cross-functional teams," he says. "No single-function team can really deliver at the speed the customer is expecting." CarMax's product teams are small, typically seven to nine people. A team can pull in staffers from any pertinent function or department, but every team includes a product manager, a lead engineer or developer, and a user experience expert, roles Mohammad calls nonnegotiable.

As we allude to in this chapter's introduction, CarMax executives provide teams with goals but not elaborate instructions. As Mohammad describes it,

> we tell them what to achieve but not how. Progress toward goals is closely monitored, with the teams expected to give 10- to 15-minute presentations every two weeks in an open-house format. The presentations address how the teams are tracking against their goals, what experiments they've done, what worked, didn't work, and what they've learned. Anyone at CarMax can attend the presentations, and top company executives often do so.

"We have this mindset of learn cheap and learn fast," Mohammad explains. "Because we are conducting two-week sprints, it's easier for the teams to conduct many experiments without adding significant risk for the business. They are encouraged to take smaller risks, learn from them, and adapt quickly." Team members are evaluated both as part of their teams and as individuals. They're expected to know and monitor their key performance indicators, and they are accountable to CarMax's CMO, CITO (Mohammad), and COO.[2]

Mohammad says the cross-functional teams were critical to the December 2018 launch of CarMax's omnichannel experience in Atlanta, a new approach to buying and selling cars that the retailer has now rolled out nationwide. The omnichannel experience empowers customers to buy a car on their own terms—online, in-store, and a seamless integration of both. He says that "basically, the omnichannel experience allows us to meet the customer wherever they are at. They can do as much or as little online as they'd like. The experience is personalized to each individual customer." If customers wish, they can now complete an entire purchase from home, with the vehicle delivered to their driveway. With the arrival of online-only competitors like carvana.com and vroom.com, and with more consumer choice in transportation options (Uber, Lyft, Zipcar, and car subscription services), CarMax is focused on retaining its industry leadership by continuously innovating its offerings.

Mohammad indicates that similar types of experimental mindsets at the team level helps the company adapt to the challenges posed by the acute COVID disruption.

> Our omnichannel approach has been critical for helping us respond to the disruption. Omnichannel empowers customers to buy a car on their terms. They can do as much or as little online as they prefer. Due to the pandemic, customers are interested in doing more of the process at home, and spending less time in stores, and the capabilities we have built over the past few years enables this. We were planning to finish the deployment of omnichannel by the end of February 2021, but because of the pandemic we prioritized it and finished its roll out nationwide in August of 2020. We also added the option for customers to buy a car online, and complete the purchase of the vehicle with a contactless curbside pickup at our stores.

Its embrace of cross-functional teams has changed CarMax's managerial mindset around technology planning. Executives and managers no longer think and talk in terms of projects and project budgets. Mohammad notes that rather than asking, "Hey, do we need X million dollars?" they say, "How many product teams are you going to fund next year and what are our business outcome goals?" He calls that a game-changer for the company: "We moved away from this annual project-based kind of mindset to a product-based mindset, where the product teams are delivering results, and we're tweaking them along the way." Product teams orient around business outcomes—like new campaigns and product launches—while project teams align around activities and milestones, which is why the product mindset is so essential to interdependent and novel work.

For CarMax, cross-functional teaming is relevant to how projects are managed and how products or offerings are developed and launched. What's common to both processes is a rapid cadence that focuses on outcomes, not detailed instructions, and is built around small experiments and tests that provide learning and the basis for scaling.

Sara Armbruster of Steelcase also emphasized the importance of an experimental mindset.

> Our culture of experimentation and prototyping has been key, even with some of the internal facing changes that we've had to make at Steelcase. It's about saying let's just try something and see how it goes. Let's talk to our employees and talk to our users and continue to iterate. Having that skill set really came in handy, and it's something that we will continue to invest in beyond the crisis.

Using Teams as an Experimentation Platform

One way that companies can engage in better decision-making is to develop organizational competencies for experimentation. Ben Waber, president and cofounder of Humanyze, is hopeful that one positive outcome of acute disruption will be an increased willingness to experiment.

> The willingness of companies to experiment and admit that something is an experiment or a test has been extremely positive. Even after the pandemic has subsided, and even if that takes well over a year, I am really hopeful that that is something that will continue. It is very healthy to be able to admit that we have a hypothesis and be free to change it based on new data that comes in.

Technology companies, like Google and Facebook, routinely engage in A/B and multivariate testing, in which the company makes small changes to one or more elements of their platforms or search results and analyze how it changes the outcomes. Richard Gingras, VP of news at Google, says that

> it's really not that important if the experiment succeeds or fails, it's what they learn from it—good, bad, or indifferent, it is intelligence that they can lay claim to. Maybe it didn't work out the way we thought it would but we learned X, Y, and Z, and we're not embarrassed by the fact that our initial assumptions were wrong. There are no failures. We tried something and we learned something.

Only by learning why an experiment was effective can you apply that learning to other aspects of the organization.

While it may be easier for technology companies to experiment extensively, the increased digitization of the workplace—particularly organizations that have shifted to remote work—generates abundant data that can enable experimentation. With such a data trail, it becomes easier to make an organizational change at one location or within one team and to assess the effect of that intervention. For example, in some prelockdown research, Waber told us that Humanyze found that simple changes to the size of lunch tables in the cafeteria at one call center massively increased the job satisfaction of individuals, because it led to more interactions between coworkers.

A lack of willingness to experiment and a fear of failure are the biggest factors holding companies back with respect to chronic digital disruption. Robert Kegan, a professor at Harvard University, argues that most leaders lack the cognitive flexibility to toggle between being disciplined and entrepreneurial.[3] Capabilities for experimental thinking enables leaders to do both. The entrepreneurial ideas help drive the hypotheses that the experiments will test, while the discipline comes from the experimental discipline. Vijay Govindarajan refers to this process as "planned opportunism."[4] Planning an experimental infrastructure to test the future enables opportunism when it is discovered through experimentation.

Managing Autonomous Teams: Loose Coupling versus Tight Controls

Organizations that excel at external collaboration and cross-functional teaming embrace organizational theorist Karl Weick's notion of "loosely coupled systems." As Weick explains in a much-cited 1976 article in *Administrative Science Quarterly*,[5] a loosely coupled organization eschews hierarchies and excels at adaptation. "If all of the elements in a large system are loosely coupled to one another, then any one element can adjust to and modify a local unique contingency without affecting the whole system," he writes. "These local adaptations can be swift, relatively economical, and substantial."

Traditional corporate teams are tightly coupled and cleanly divided. Staffers in, say, information technology work closely together and let their technical expertise guide their work. When they need to tap into the knowledge of other departments, they do so through formal channels, asking their manager to confer with her counterpart in, say, accounting or marketing. A cross-functional team, by contrast, is loosely coupled in that it is composed of people from multiple functional areas with the freedom to work across traditional organizational boundaries.

Digitally maturing companies embrace loosely coupled relationships, systems, and processes to support their digital innovation. They give greater autonomy to their cross-functional teams and individual units, which have the freedom to respond quickly to shifts in their market environment. Yet, these teams can't be operating independently. Michael Aldridge of Humana notes that "it involves more telling senior leadership what we're doing to keep them in the loop and let them know how they can support us, rather than asking for permission to act."

Their interactions with external partners are governed more by relationships than by detailed contracts. These stronger relationships enable the cross-pollination of skill sets and mindsets, which in turn allows novel solutions to arise more often and more quickly than in tightly controlled systems, meaning the overall system is less vulnerable to the breakdown of any one part. Increased autonomy does require different forms of governance: it demands sturdy ethical guardrails to ensure that the autonomous units serve the firm's overall goals and to protect its reputation. We address aspects of these guardrails in chapter 3 on purpose, values, and mission.

How to Apply the Concepts from This Chapter: A Checklist for Teaming

The hallmark of agile organizations is teams. Agile organizations value high-performing teams over high-performing individuals and look to teams to unlock the value of individuals. High-performing teams are inevitably worth more than the sum of their parts. For a culture of teaming to truly flourish, we believe that teams need to meet three broad conditions: (1) a shared goal, (2) strong feedback mechanisms, and (3) psychological safety.

In today's disruptive world, organizations need new ways to communicate and collaborate—through informal structures, networks of relationships, and customer-oriented outcomes. We recommend organizations employ three types of teams, depending on the mission they serve:

- **Shared services**: High-efficiency teams that focus on operational tasks that are standardized and transactional and with well-defined interfaces with the rest of the business
- **Resource pools**: Highly specialized workers who temporarily join other teams to provide expertise, as demand for their time is high across the entire organization, but are not constantly in specific areas within the organization (e.g., data scientists)
- **Project teams**: Composed of individuals from multiple disciplines who are adaptable, mission centric, autonomous, and collaborative. These teams are most often organized around a specific problem or target outcome.

Nimble organizations are those that can identify the right type of team for the outcome they need, bring together individuals, and disband them when their mission is complete. Indeed, these organizations rethink the way they make decisions and adapt using a test-and-learn approach.

The Team
Use the following questions/checklist to determine whether you've built a high-performing team or one that isn't set up to succeed.

- Does the team have a clear, measurable **goal**?
- Is the goal centered around the **customer** (internal or external) and their interests?
- Are there clear **feedback** mechanisms in place to quickly self-correct (e.g., metrics)?

- Is the team being objective and **data driven** with all decision-making?
- Does everyone have equal **influence** over outcomes (e.g., no dominating personalities)?
- Do team members intrinsically **trust** each other and the team leader (i.e., psychological safety)?
- Is there **accountability** for team outcomes (e.g., no individual successes and failures)?
- Are there mechanisms in place to **recognize** the team and drive engagement?

If you've checked off more characteristics than you left unchecked, you are likely already doing better than a lot of other readers. However, we would encourage you to continue building an environment where teams can thrive (check the other boxes) by partnering with team leaders (see below) to help close these gaps. While services-led organizations have traditionally been built around empowered teams, high-tech companies arguably do this better than anyone else. Think about the last time you tried to create a team—whether a single team or as part of a broader organizational transformation; did it succeed or fail? If it failed, was it due to one of the reasons above?

The Leader

Nimble organizations require a new definition of leadership. Traditional leaders focus on ideation, instruction, and inspiration. However, increasingly, team leaders throughout all levels of the organization must be orchestrators who foster an environment in which high-performing teams can succeed. Specifically, leaders must

- energize the team, including cocreating the team's missions and goals and providing a psychological safety net;
- empower the team, allowing teams to make their own decisions and cultivating the skills and behaviors required to succeed; and
- connect the team(s), building intentional collaborations and facilitating buy-in from stakeholders, as required.

These types of leaders are hard to come by, though fortunately they can be developed. There's no one way to build a checklist around leaders. Depending on the goals of the enterprise, type of team, and intended outcome, specific skills might change. Nevertheless, we have seen through research—time and time again—that these eight skills are evergreen:

- Social **flexibility** (e.g., ability to work with people from different backgrounds)
- **Regard** for people (e.g., ability to be empathetic toward and respectful of opinions)
- **Self-drive** (e.g., intrinsically motivated)
- Willingness to **experiment** and take risks
- **Resilience** in the face of change, disruption, and challenges
- **Decisiveness**, especially in tough moments and (sometimes) without all the information
- **Conceptual** thinking (e.g., ability to establish and drive toward an ambiguous goal)
- Breadth of **perspective** (e.g., ability to draw on a variety of experiences)

Consider the list above a self-evaluation (if looking inward) or a way to test someone's readiness for leadership roles (if looking outward). Checking all eight boxes is difficult—don't be disheartened if you or someone else falls short. Instead, use this as an opportunity to celebrate your strengths, and build from these strengths to develop new ones. We've often said that this is a golden age of learning. There are numerous ways to build skills, from courses to mentoring and coaching (see table 9.1). No wonder then, that these leaders will be among the hardest for the technological revolution (read: AI) to replace. So, will you be a leader who transforms your team, your organization, or even your ecosystem into an agile, customer-centric, goal-oriented, and self-learning organism?

11 Rebuilding Disrupted Customer Relationships

Our lives are largely determined by factors we never fully notice: our habits, those unthinking, automatic choices that surround us each day. They guide how we get dressed in the morning and fall asleep at night. They affect what we eat, how we do business, and whether we exercise or have a beer after work.[1]

—Charles Duhigg

The greatness of humanity is not in being human, but in being humane.

—Mahatma Gandhi

For most of this book, we have focused on how companies can adapt their strategies and operations to respond to a disrupted environment. This chapter focuses on another aspect of disruption: the profound impact on customer behaviors, preferences, and interactions with brands—and what organizations can do to rethink their interactions with customers. While we save a discussion of customers for last, it is by no means least. A recent study by Deloitte shows that leading companies are more likely to prioritize a focus on customers, use technology to improve their customer experience, and even engage their customers in innovation.[2] But first we need to understand the role of habits in shaping consumer behavior, as habits are far more likely to change quickly and often without warning during an acute disruption than chronic ones. The rapid disruption of habits may be one of the most significant distinctions between chronic and acute disruption. You can maintain habits in the face of chronic disruption or adapt gradually but you often cannot with acute disruptions.

Habits Die Hard . . . until a Pandemic Strikes

In *Atomic Habits*, James Clear shares research indicating that between 40 and 50 percent of an individual's actions and decisions are habit driven.[3] As Clear and Charles Duhigg, author of *The Power of Habit*,[4] argue, habits are automatic, not necessarily conscious, choices that influence many of the small and big daily decisions that we make.

Clear notes that people generally like to believe they are in control of their actions. The reality is, however, that many of our daily actions result not from conscious choice but simply from selecting the most apparent option driven by our experience. Habit means that people are seeking to derive the greatest value for the least amount of effort. People generally act in ways that are easy. Clear recommends that people redesign their habits such that the decisions that provide the most value are also the easiest to choose. Technology can be a powerful tool for automating and reinforcing positive habits, which is perhaps the most effective way to engage in desirable behavior.

Habits influence not only what we eat, how much we exercise, and how we spend our free time but also our behavior as customers. Where we choose to shop, when we choose to shop, and whether we pick a tried-and-true brand versus a product on sale are all determined by habit. There are times when customers engage and stick with their companies or brands, not out of loyalty but out of habit. In early 2017, *Harvard Business Review*, published an article by former Procter & Gamble CEO, A. G. Lafley, and coauthor Roger L. Martin, provocatively titled "Customer Loyalty Is Overrated: Focus on Habit Instead."[5] Americus Reed, professor of marketing at the University of Pennsylvania's Wharton School, refers to this as "consumer inertia."[6]

Think about this the next time you choose a product in a particular category. Do you examine every single type of laundry detergent on the grocery shelf or immediately lock on the detergent in the familiar red plastic container? Or perhaps you just have this purchase as an auto-replenishment option on your online shopping service. Why? Habit, that's why.

If habits trump conscious choices in determining much of our engagement with brands, a disruption to those habits—especially significant ones associated with acute disruptions—can usher in a massive disruption of habitual shopping and buying patterns. It forces customers to find new ways to solve problems and meet their needs. Customers' habitual ways of doing things either may simply not be available anymore (e.g., live theater and sporting events cancelled), or they may involve much higher real (or perceived) costs than before the disruption (e.g., brick-and-mortar grocery shopping are seen as creating a higher risk of infection).

Data from the early days of the acute COVID disruption suggests the potential for significant shifts in such behaviors.[7] A survey of shoppers in late March 2020 reveals that 85 percent shopped at a physical store that was new to them following the lockdown and 75 percent shopped at a new online location. Fifty-four percent indicate that they had purchased products from brands that were new to them. This phenomenon is not confined to business-to-consumer (B2C) customers; B2B customers also expressed significantly higher preferences for digital interactions. Survey respondents indicated the likelihood of choosing a company that provided an outstanding digital experience as a primary supplier doubled during the COVID disruption.[8]

Even though new habits normally form over an extended period, acute disruptions can shock many customers into adopting new habits almost overnight. Research from Deloitte further suggests that many of the behaviors developed during the lockdown are here to stay for both B2C and B2B customers.[9] Personal experience supports this finding. We, along with many of our friends and colleagues, have assiduously avoided visits to grocery stores and other major brick-and-mortar locations (e.g., banks) since the lockdown began and have no immediate plans to resume our prior behaviors. We cannot believe that we are the only ones.

As a result of the pandemic, habit disruption is happening at scale versus one customer or one segment at a time. Millions, if not billions, of customers globally have had habits disrupted, and this manifests in new customer behaviors. This disruption represents both a massive threat and opportunity for organizations. If the habits that your customers rely on for their interactions with your company are disrupted, it raises the possibility that customer engagement patterns will change radically and possibly permanently. They may leave your brand for a competitor or a substitute product, or they may simply decide they do not need your offerings after all.

This challenge of disrupted habits at scale presents two strategic considerations for companies. First, if your customers' habits are disrupted, what actions can you take to try to preserve those relationships amid the disruption? Second, what opportunities does a disruption in your competitor's customers' habits present for your company, and how do you take advantage of those?

Keeping Customers by Establishing Trust

"Trust but verify" is the English translation of a Russian proverb that became famous when former US president Ronald Reagan adopted and regularly used it in the context of nuclear disarmament discussions with the Soviet Union. Like Reagan, who was looking for proof that the Soviet Union was willing to disarm, customers who are in the midst of their own pivots are looking for assurances that companies can deliver what they need in a disrupted world. The shock that customers experienced to their habits may present an opportunity, but as the saying goes, you may only get one bite of the apple.

Eric Ranta of Google Cloud notes the importance of strong customer relationships to help retain customers during the disruption caused by the pandemic. He observes that "if you look at companies that really thrived out of this situation, they had a service business model where they were always thinking about what their customers needed and what they represent. They've been able to perform fairly well because they've listened to the customers."

Trust in customer service is probably one of the hardest hit functions in the case of a pandemic. Matthew Dixon, Ted McKenna, and Gerardo de la O "used a proprietary 250-variable algorithm that . . . score[d] the effort level of a customer's interaction—ranging from 'difficult' to 'easy' for the customer to accomplish their goal—as well as the underlying drivers behind those scores . . . [for] roughly 1 million customer-service calls involving more than 20 companies representing a broad cross-section of industries."[10] They found that "the average company in our study saw the percentage of calls scored as 'difficult' more than double from a typical level of 10% to more than 20%."[11]

As you try to skate to where the puck is going to be (to appropriate the spirit, if not the exact words, of the hockey great, Wayne Gretzky), you need to focus on establishing trust with your customers across four dimensions: humanity, transparency, capability, and reliability. Of these, we found humanity to be the most salient.

Humanity: Competing on Caring

Humanity is about whether an organization genuinely cares for the experience and well-being of others. Expressing humanity includes concerns around safety (prioritizing people's health and security), respectful treatment (valuing and treating all customers, employees, and partners with respect), and an attention to the greater good (balancing individual well-being, environmental, and societal interests with profit motives).

Safety

During the acute COVID disruption, the importance of safety came to the forefront. Shamim Mohammad of CarMax notes the rise of safety and security as a distinct customer need.

> What's different is that now customers are also looking for more security and safety. Those two things really have changed. We worked quickly to put

measures in place that encouraged safety and security for both our customers and our associates, so we could continue to deliver unmatched customer
experience, even during a pandemic. We developed and deployed contactless curbside pickup in just three weeks and also began offering solo test
drives, so our customers could test drive vehicles without coming in close
contact with our associates. I'm really proud of the agility and innovation
our teams applied during this time.

A big part of addressing safety is modifying physical interactions to
increase customers' trust in physical locations. Robert Shumsky and
Laurens Debo suggest four strategies for building this trust.[12] First,
make customers feel safe while shopping by de-densifying the location,
and encourage ways to make the shopping experience more efficient
to increase customer throughput to minimize impact on profitability. Second, manage variability in the context of physical location.
Make sure customers know what to expect when they visit you. Third,
develop "nudges" to encourage safe behaviors. You can't just tell people
how to behave; you need to show them too. Returning to the work of
James Clear, you are trying to shape new habits, and these new habits
need considerable environmental "scaffolding" to take root. Fourth,
emphasize why you are shaping the space in ways that you are, to both
improve the customer experience in the new environment and provide
for the safety of your employees.

While many of these procedures may be specific to the COVID disruption, the underlying message of needing to create a physical space and
practices that prioritize human experience and well-being will persist.

Respectful Treatment

Times of acute disruption and high uncertainty can often create fear
and stress—and in some cases an overwhelming sense of grief over
loss of normalcy.[13] During the acute COVID disruption, matters were
made worse because people were unable to take comfort by coming
together socially in groups and communities the way they had before.
The disruptions of 2020 are wreaking havoc on our individual and collective mental, emotional, and physical beings. We yearn for stability,

reliability, things we can trust, and things that make us feel connected. In response, companies can help by supporting genuine human connection through respectful and supportive treatment. Emma Lewis of Shell emphasizes this point, saying, "People always remember how you treated them in a crisis, whether it's an employee or a customer. I think our crisis management has strengthened those bonds with what we saw as our target customer, and it will actually change our segmentation and services in the future."

Before 2020, US Surgeon General Vivek Murthy called for our attention to a public health crisis: a silent epidemic called loneliness. When his book *Together: The Healing Power of Human Connection in a Sometimes Lonely World* came out in the spring of 2020, it was timely, as many people were being driven apart physically and often unable to connect with family, friends, and communities in ways they used to. As humans, we are social creatures and need to feel connected, and yet all of the different ways we connect to each other were being diminished. Even a simple exchange in a grocery store or a restaurant is now very transactional and designed for efficiency rather than for human connection.

Yet this emphasis on health, safety and social distance also risks the chance of losing genuine human connection. As Brian King of Marriott states, "Contactless is an interesting conundrum for hotel professionals since the core of the hospitality business is connecting to customers. . . . What does it mean to be hospitable in a contactless environment? You must find a way to make digital hospitable."

Customer service representatives need to be trained in a way that communicates the compassion and caring the company seeks to cultivate with the customer. Grant Packard, Sarah G. Moore, and Brent McFerran suggest several ways businesses can communicate that increase trust.[14] The first is using concrete and specific language. They note a greater specificity in language helps in times of uncertainty. For example, a barista saying "can I get you a coffee?" instead of "how can I help you?" is more specific, concrete, and tailored to the situation. Customer-facing employees should tailor language wherever possible to a customer's unique and specific needs to show they are listening.

Second, they suggest employees can build trust with customers through relationship building. Employees should always use "I" and speak on behalf of the employee rather than a "we" on behalf of the company. People are more willing to trust individuals than companies and policies. You also need to empower these employees to be trustworthy by allowing them to advocate on behalf of the customer. Third, they encourage companies to not just be competent but also caring—customers want both. While the bulk of the exchange should reflect the employees' competence, the beginning and end of the conversation should be a time when the employee can convey warmth. This communication pattern helps not only solve the customer's problem but also lets them know they are valued as a customer.

Greater Good

In times of disruption, it is essential that customers do not think the company is simply out to support its own financial interests. Particularly in times of disruption, when people are feeling the pressure of uncertainty and fear, showing that your company is seeking to help on a broader level is an important way to establish trust.

Doug Mack, CEO of the sportswear retailer Fanatics, provides a story of pivoting operations to create personal protective equipment at the peak of the acute disruption.

> There was a huge crisis with personal protective equipment, where medical workers couldn't get masks. You have these incredible heroes and healthcare workers trying to take care of patients who are actually putting themselves at risk with ad hoc protections. So, we have a domestic manufacturing facility in eastern Pennsylvania that got the Governor's approval to operate. We converted that to a mask and gown manufacturing facility literally overnight. We produced face coverings and gowns out of the material we normally use for Major League Baseball uniforms. We actually sent that material to one of the hospitals to run through testing protocols to make sure it was good material for the purposes. We ended up manufacturing over a million masks and gowns, donating those to hospitals and frontline workers in 13 states.

Matt Schuyler of Hilton also emphasizes the greater good by describing the company's efforts at opening rooms to frontline healthcare workers in New York and New Jersey at the peak of the surge in April 2020. He explains the impact on this decision both on company morale and customer perception: "We were recognizing that healthcare workers in the first responders who were exposed to infections didn't want to go home to their families and potentially infect them. Along comes Hilton to say, we'll put you up between your shifts. We know you need some rest, so we'll put you up. It created a huge sense of pride internally."

Noah Glass of Olo emphasizes the importance of prioritizing the greater good to underscore the focus on humanity: "We very much buy into the idea of conscious capitalism or enlightened hospitality. The employees come first, then the customers, then the partners in the community, and only then the investors. That's the stack ranking of stakeholders to the business. That was our guiding principle during this moment of crisis."

Michael Aldridge of Humana notes a final example of a company responding to the greater good. Humana took two floors of an office space primarily used for employees who could readily work remotely that was scheduled for renovation. They postponed the scheduled renovation to repurpose the space to use for in-person schooling for low-income and at-risk kids, essentially expanding the physical plant of the local public school system.

Transparency, Capability, and Reliability

While an emphasis on humanity is paramount in acute disruption, other factors are also essential for companies seeking to retain relationships with customers—transparency, capability, and reliability.

Transparency

Transparency represents the degree to which an organization openly shares information, motives, and choices in plain language. This includes honest and accurate communication, unmistakable motives, and straightforward language that is plain and easy to understand. In an age of big data, companies can build trust with customers by creating more transparency in their collection, use, and management of customer data. At this point in time, most of us are pretty skeptical about providing data to companies since we believe that most of them will simply use that data to try to sell more goods and services to us. But customer data can also be invaluable in helping customers navigate through complex product choices and configurations.

Many of our respondents indicated the value of transparency in communication both inside and outside the company. For example, Kristin Darby of Envision Healthcare noted that "leading through a crisis requires clear, consistent and frequent communications. Being transparent means delivering information in a way that people can absorb what you know and what you don't know and how it may impact them."

Matt Schuyler describes the role of transparent communication for Hilton.

> The backdrop of the entire process was to be honest and direct through an active communication strategy. We were as transparent as we, I believe, possibly could be with respect to the impact the crisis was having on the company. We frequently hear an appreciation of the transparency and the thanks on doing everything we could to help an impacted workforce.

Monty Hamilton, CEO of the digital engineering company Rural Sourcing, also emphasizes the role of transparency: "We made sure that we're communicating everything that we do and what we know, which wasn't a ton of the early stages. We're trying to be as transparent as we possibly can to make sure that these folks know we're making the best decisions on the limited amount of data we have."

Brad Keller of Humana indicates some of the ways communications with customers and stakeholders would likely change following disruption: "There's going to be a need for increased transparency of cleaning routines. Cleaning workspaces is something that goes on behind the scenes after hours that we don't know about. At most, you might see one of those little letter-sized pieces of paper in the bathroom door that says when it was cleaned. And now we demand to see it. We want to know."

Organizations that offer AI-enabled recommender systems find that customers appreciate online recommendations divorced from high-pressure sales tactics. An industrial reseller posted a maintenance and repair content recommender, including video, on its site that bundled product and service advice and allowed curated comments on the site. Very quickly, the site generated fifty times more web traffic than the reseller's biggest supplier, presumably because the customers valued the utility of how the site combined their information and product information to provide valuable recommendations.[15]

Capability and Reliability

While humanity and transparency signal the intent of the organization, capability and reliability form the foundation of the organization's competence. Capability is the degree to which an organization possesses the means to meet expectations, while reliability is how much an organization consistently and dependably delivers on promises made. This includes the safety, quality, value, and dependability of the products and services.

The chief digital officer of a large, global insurer underscored the importance of reliability: "We run a series of strategy sessions and we ask people, what are the words you'd use to describe the company. The words that came up were stable, safe, reliable, confident, trusting—all the sorts of words that matter so deeply at the moment. Six months ago, they likely would have said something different."

Nowhere was the issue of capability and reliably clearer than in the supply chain disruption that many companies experienced. In 2020, with broken supply chains, much of the reliability of products and services came into question, along with the capability of many organizations to continue to serve under new conditions. The acute disruption shifted demand so significantly that even robust companies had difficulty maintaining operational standards. For example, delivery companies had difficulty meeting the demand to deliver packages in areas like Massachusetts's Cape Cod, because the uptick in online shopping combined with the influx of people arriving early to vacation homes from cities like New York and Boston completely overwhelmed capacity. Toilet paper, disinfectant wipes, and hand sanitizer were virtually impossible to find in the early days, followed by home grooming equipment and certain food items, and then bicycles and even certain models of automobiles. Numerous respondents reported the need for reliable supply chains as a significant problem through the disruption. It's difficult for your customers to trust you when you can't provide their products in a timely manner—or at all.

Eric Ranta of Google Cloud describes his own experience with disrupted supply chains and how he envisions it changing customer behavior.

> I don't think we fully appreciate how supply chains will change, and the consumer behavior that will come out of it. There were a lot of dark days in the world. We thought our food supply chain is going to go away for a while, and now its cars or certain kinds of toys—a wave of these things that are still working through the system. Customer decisions will be different as a result of these experiences.

Competing for Disrupted Habits

Habit disruption can certainly threaten your existing customer base. On the other hand, this habit disruption represents an opportunity because the habits that your competitors' customers rely on to interact with them are also massively disrupted, creating new opportunities to engage with these customers. For example, sales of Peloton bikes jumped 66 percent during the early days of the pandemic. Will this continue when face-to-face training resumes?[16]

Brands have always sought to use disruptions as an opportunity to win new customers. As Geoffrey Moore chronicles in his classic book, *Crossing the Chasm*, Salesforce took advantage of the disruption created by cloud technology to scoop up customers with its SaaS offering that had previously been unable to benefit from the customer relationship management software offered by software giants.[17] Other SaaS providers quickly followed suit. To use a metaphor shared by Noah Glass of Olo, a provider of SaaS platforms for multilocation restaurant businesses, habit disruption is like the ground ball in lacrosse where players chase after it at the same time.

Chris Whitfield of McDonald's notes the potential with a shift in customer habits. He says that they've experienced substantial changes that just come from safety considerations and changing customer habits. We're recognizing that this is a window where we have potential to invest in a way that sets us apart from our competitors. Maybe it doesn't set us apart tomorrow, because it takes some time to implement these things. If we start that process now, though, we might come out the other side of COVID -- with our investments having strengthened our position.

Acute disruptions present unique opportunities to identify and market to the new habits customers are rapidly forming in real time. Of course, because customer habits are changing in unexpected ways so quickly, data that we are accustomed to using to predict customer behavior may be of limited value in helping us understand how to win the ground ball. Daniel Leibholz of Analog Devices notes that "our customers are changing rapidly. They are more software-centric, and they expect our products will be high performance and easy to use. We are creating new ways for customers to evaluate and select our technology through digitalized environments. We are using those environments to continuously evolve our approaches to engaging with customers."

Fortunately, Michael Jacobides and Martin Reeves describe a two-step process for identifying the types of opportunities that may be present following a disruption.[18] First, they recommend examining the cascade effects that a change in habits is likely to bring about. For example, they explore in depth that the shift of more time at home will have on several activities, such as working, dining, and entertainment. They then plot which trends will likely increase as a result of this behavioral change (e.g., home office materials, takeout dining, in-home entertainment) and which will likely decrease (e.g., business travel, dining out, live entertainment).

Second, they plot the nature of these changes on a 2×2 matrix, assessing whether the change is likely to be temporary or permanent and whether it represents a new or existing trend. They classify the results into four different types of shifts:

1. **Boost**: A temporary acceleration of existing trends. Video game play probably falls into this category. Video game play increased substantially during the acute COVID disruption, but it may be a short-term

trend that will revert to more normal usage levels once people can return to other activities.

2. **Catalyst**: A lasting acceleration of preexisting trends. In contrast to video game play, video streaming services may fall into this category. The competitive environment was already shifting toward streaming services before the acute COVID disruption, and the disruption likely amplified the impact of these preexisting trends. We do not see customers abandoning streaming platforms en masse once they have been adopted. Online shopping also likely fits into this category.

3. **Displacement**: A temporary shift toward new trends. Jacobides and Reeves refer to the uptick in jigsaw puzzles as an example of this category. People adopted new behaviors to deal with the situation of the lockdown, but they are unlikely to sustain themselves once the disruption is over.

4. **Innovation**: A permanent introduction of a new trend. Jacobides and Reeves place direct-to-consumer theatrical releases in this category. Uncertain when people will return to the movie theater, studios need new avenues for reaching customers. If studios can identify productive business models in this category, they may never return to the in-person movie theater.

Once you understand the nature of the four different types of behavioral shifts, you can then begin to consider how to adapt your business to the new behaviors. As our colleague Larry Keeley and his coauthors discuss in *Ten Types of Innovation: The Discipline of Building Breakthroughs*, innovation can assume many different guises beyond glitzy new products, ranging from new experiences to new forms of customer service to new pricing models to new ways to link your offerings with offerings from others.[19] For example, the Chinese cosmetics company Lin Qingxuan turned its in-store beauty advisors into online influencers, helping shift sales to online channels.

Understanding habits and using technology to effectively pivot when acute disruptions cause habits to shift is key for capturing new customer relationships during disruption. Yet, it's also critical to build trust in those new and existing relationships—through humanity, transparency, capability, and reliability in order to create amazing customer experiences.

How to Apply the Concepts from This Chapter: A Framework for Customer Engagement

Table stakes needs for customers fall into two categories:

1. **Better performance or pricing**: Offer higher performance (e.g., Porsche) and/or a better price to value ratio (e.g., Target).

2. **Compelling experience**: Ensure customers have a positive and pleasant journey across the learn, buy, get, use, pay, and service life cycle to gain "sticky" followership.

Most companies do—or try to do—both of these well. However, the basis of competition has changed. Enterprises can also offer improved functionality and deeper connections to strengthen the relationship with the customer and reinforce signals of humanity, transparency, capability, and reliability discussed in this chapter.

This framework provides an approach for how companies can elevate themselves beyond the "table stakes" value they provide customers and overcome customer inertia. Here are seven potential strategies adopted from the book *Ten Types of Innovation: The Discipline of Building Breakthroughs*.

Improving functionality (including capability, reliability):

- **Reduce effort**: Simplify jobs to be done with products or offerings that "lighten the load" for customers (e.g., Handy, Uber Eats).

- **Provide guarantees**: Reduce any uncertainty around the acquisition and use of an offering (e.g., trial periods, freemium business models).

- **Enable customization**: Provide flexibility to allow customers to make the core offering their own to meet individual preferences (e.g., Nike, Starbucks).

Table 11.1
Applicable tactics for strengthening customer relationships

Symptoms/market dynamic	Applicable strategies
Is my company difficult to do business with?	• Reduce effort • Magnify purpose • Become a trusted advisor
Is choosing to do business with my company risky?	• Provide guarantees • Enable customization • Build identity
Are my current offerings relatively undistinguished and commoditized?	• Enable customization • Surprise and delight • Build identity
Are my current offerings centered around utility?	• Provide guarantees • Enable customization
Is it difficult for my customers to achieve their goals using my offerings?	• Reduce effort • Magnify purpose • Become a trusted advisor
Do my offerings require specialized skills or knowledge to use effectively?	• Provide guarantees • Magnify purpose • Become a trusted advisor

Deepening connections (including humanity, transparency):

- **Surprise and delight**: Deliver memorable experiences, based on differentiated capabilities, to meet unexpressed customer needs and desires (e.g., Disney, Marriott).

- **Build identity**: Establish a coherent identity that resonates with customer values and instills a sense of group affiliation and belonging (e.g., Apple, Warby Parker).

- **Magnify purpose**: Create tools for users to achieve their goals, and facilitate self-improvement and self-actualization (e.g., Fitbit, Charles Schwab).

- **Become a trusted advisor**: Establish a relationship as a trusted partner and key source of input and counsel when making decisions (e.g., Best Buy, Home Depot).

Which symptoms suggest opportunities to employ these approaches? Table 11.1 covers some symptoms and applicable tactics.

Epilogue: Is the End Near?

How Did We Get Here?

Writing this book has been a unique experience. We had actually prepared a pitch for a very different book when we first submitted our proposal to the MIT Press in early March 2020. Yet, between the time of our initial proposal and our meeting with the editors, it became apparent that a world-altering disruption was beginning to take shape. It also became clear that our previous research on digital disruption was going to be particularly relevant to the challenges that most companies were beginning to experience. So, we pivoted during the pitch meeting and proposed researching and writing a book about how companies were navigating the acute disruption that was just beginning as a result of the COVID-19 pandemic.

The process of researching and writing this book while also living through the disruption ourselves was stressful. Not a week passed where we didn't wonder what we had gotten ourselves into and whether we could actually pull off the goal we had signed a contract to achieve. It was painful to hear some leaders talk about having to lay off thousands of employees and others talk about figuring out how to save businesses with underlying business models that disappeared overnight. The process was extraordinarily fluid, as we investigated a phenomenon that evolved and unfolded before our eyes. Throughout most of the process, it wasn't clear how severe it would be or how long it would last. How could we write about a phenomenon when so much was still unknown? When would we know enough?

As our research proceeded, we felt as if we had a front row seat for likely the most significant event that most organizations would experience in their lifetimes. We appreciate the candor and honesty that all of our interview subjects exhibited. We were surprised at the number of leaders who viewed the disruption as an opportunity to move their organization forward. Some found unprecedented success as their business or function became critically important to the emerging environment. It was also gratifying to talk to so many business leaders who were clearly stepping up, thinking creatively, and acting boldly to lead their organizations. Almost every interview revealed a surprising insight that triggered new ideas and perspectives. Not a week passed where we weren't fascinated and inspired by the stories we uncovered. We are proud of the work we completed, and we hope it will endure long after the acute COVID disruption is behind us.

Timing Is Everything

Throughout the project, we have obsessed about the timing of the book's release. Put simply, there is a long lead time between the submission of a book manuscript and when it finally appears in print. We appreciate the flexibility of our editors at the MIT Press with this process as much as possible without compromising standards of quality.

Nevertheless, our concerns lingered. Would we publish the book too late, with the phenomenon of the pandemic long over and a distant memory of little interest to readers? Might we publish too early and end up missing major developments during this period of disruption? Would something change significantly between the time we submitted our manuscript and when the book appeared in print, making our insights seem out of touch, dated, or simply wrong?

In an effort to mitigate some of these risks about timing, we asked for an extension to write this epilogue—and we are glad we did! Between the time we submitted the manuscript and writing this epilogue, two major events occurred that made the timeline of the intervening months clearer.

The first is the outcome of the US elections, heralding a change in administration. While there was an acute political disruption following the election and it seems that chronic political disruption and dysfunction in the United States may continue for the foreseeable future, the direction of governance is clearer than it was when we submitted the manuscript. Several of our interviewees lamented the lack of stronger and clearer leadership from the US federal government in the face of the pandemic, and we hope a change in administration will provide clearer leadership to confront the challenges that lie ahead.

The second is the announcement of several highly effective vaccines. The first of these vaccines have just been approved as we write this epilogue, and experts expect them to be widely accessible by the time our book is published in early fall 2021. Although there remains some concern about the vaccines being less effective against emerging variants, headlines suggest a light at the end of the tunnel despite the possibility of a number of challenging months to endure before that light becomes reality. Nevertheless, we now have an approximate timeline regarding when the acute COVID disruption is likely to be over, and it seems that this book will be released as we embark upon that "next normal."

The immediate postpandemic environment creates the perfect audience for our core message. As we slowly emerge from this disruption, we expect that many companies, leaders, and employees will face a strong temptation to slip back into the comfortable and familiar routines of the prepandemic workplace. If so, they risk repeating the experience that almost beset KLM in chapter 2, breathing a sigh of relief when the disruption is over, returning to the way things used to be, and failing to capitalize on the innovations and advances they had made during a crisis.

Widespread Vaccination Is Itself a Disruption

Like the COVID-19 pandemic, the widespread distribution of the COVID vaccine is a major disruption in and of itself. We strongly believe that there is no "going back" to the prepandemic workplace. The time following the onset of the pandemic has been a period of incredible innovation, often inspired by desperation. Organizations and individuals have had to discover new ways of working because—put bluntly—they had no choice. Many reported successfully implementing years' worth of digital transformation plans over the course of a few months. Whether it is CarMax accelerating a completely contactless auto purchasing process, Hitachi adapting sensors to monitor social distancing in factories, Freddie Mac implementing remote building inspection, and countless healthcare providers pivoting rapidly to telemedicine, innovation in response to real and immediate existential threats is a key theme in our research.

Many of the changes that organizations made in response to the pandemic were ones they needed to make anyway, as aspects of the traditional business environment were already becoming obsolete and outdated before the pandemic. The acute COVID disruption did not lead most organizations to do fundamentally *different* things then they were already contemplating; it simply forced action and transformation with urgency and at scale. Transformation success during the pandemic has fueled new confidence and optimism about our ability to harness disruption. As a result, the competitive environment that organizations emerge into will be radically different—with more nimble, scalable, and stable competitors, who have many more options to leverage.

We argue that leaders should also apply the core lesson of the book to the disruption caused by widespread vaccination that allows the safe return to public spaces—innovate through it. The postpandemic world will be fundamentally different than the one that existed before or during the acute COVID disruption. It is in precisely this way we hope readers will use this book. The tools we provide in each chapter can help organizations plot their path forward to address this next disruption. Take the following examples:

- All companies can benefit from engaging in the scenario planning process we outline in chapters 1 and 4, identifying their strategic capabilities described in chapter 2 and engaging in frank discussions using the questions we provide in chapter 3 about purpose, values, and mission.

• Regardless of the competitive environment, it is likely that nimbleness, scalability, stability, and optionality will be valuable capabilities for all organizations going forward. These capabilities simply make organizations into fiercer competitors.

• We expect that moving organizations to cloud-based infrastructure, more data-driven decision-making, and more robust security features are likely existing trends that were simply accelerated by the acute COVID disruption, not created by it.

One of the most profound decisions companies are likely to face is how and to what degree to return to colocated work. All companies will need to fundamentally rethink the workplace and discover how to leverage the best of both colocated and virtual channels across the organization, within teams, and with customers to create the best possible experience for all parties. If you do not do it, it is almost certain that your competitors will. The habits we have all adopted amid disruption are not likely to wane overnight.

The good news is that early data from Asia suggests a little colocation can go a long way toward reducing the limitations of remote work. The workplace analytics firm Humanyze learned that employees who return to the office only one to two days per week increased the number of serendipitous connections by about 25 percent, with little reduction of the performance advantages that these companies experienced with remote work. Yet, we also expect that decisions about balancing remote and colocated work affect and can be affected by a broader set of societal factors, such as access to home office space or public transportation load, all of which have been affected by the pandemic. Most companies should be thinking about how to achieve the right balance between virtual and colocated work, not deciding between one form or the other. Likewise, companies should also avoid categorizing employees as in-person and remote, as this approach is not likely to achieve the potential benefits of a hybrid work environment.

How to Use This Book for the Future

If you don't read this book in the immediate postpandemic environment, however, it is not too late. We expect the postpandemic disruption to be closer to a chronic disruption than an acute one, unfolding over many years. That doesn't mean you can afford to wait since, as we have noted, many organizations were already behind with respect to responding to chronic digital disruption. It just means there is still time to move your organization forward and innovate with what you learned. After all, KLM waited several months before it began capitalizing on the innovations established after the volcano eruption, and that was a much more short-lived disruption. It's likely not too late for your organization either.

Of course, consistent with the themes of this book, it is highly possible that the timeline unfolds differently over the coming months than we predict today. After all, we have lived through multiple major acute disruptions over the course of 2020 and early 2021; why should we be surprised if another one happens unexpectedly? It would be foolish of us to predict *how* it might proceed differently, but we can still recognize that it might. Whether companies are dealing with the predicted acute disruptions or an unexpected one, the lessons in this book still apply and the tools we provide should prove useful.

We have sought to make our findings relevant to disruption in general. Just as many of our interview subjects indicated that they looked to previous disruptions of the 2008 financial crisis or the September 11 attacks in 2001 for insights about how to lead amid the acute COVID disruption, we hope this book provides a perspective that others can use to provide insights to help leaders navigate whatever disruption, large or small, they face next.

We also hope to use this uncertain future to underscore one of the central themes of the book—disruption is here to stay and transformation efforts consequently need to be refreshed continuously. In many ways COVID-19 has also been a stress test for future disruption. As acute and chronic changes in the business environment accelerate in an overlapping fashion, winning companies will be those that continually adapt to and thrive in this emerging reality. Although we anticipate that we will be emerging from the pandemic when this book is published, many of the other disruptions we've experienced—racial inequality, climate change, digital disruption—will remain.

We hope that future readers can look to this playbook for stories and examples for how to respond to these disruptions, whatever they may be. The only response to a world characterized by disruption is to build the capabilities to engage in ongoing workplace reinvention. We are reminded of the words from Shamim Mohammad of CarMax that we quote in our introduction. None of us can predict what the future will look like; we can only prepare our organization to adapt as it comes into view and adapt again as that view shifts. This mindset proved invaluable in the pandemic. It will be invaluable after the pandemic too.

The Crucible of Leadership

Consistent with our perspectives in chapter 3, the acute COVID disruption has been a leadership crucible for many individuals across a broad set of organizations. We have seen corporate leaders emerge as strong advocates for their employees, customers, and communities, but we have also seen frontline managers at the forefront of innovation and adaptation. We hope leaders keep many of the characteristics that have helped them navigate this crisis—empathy, authenticity, experimentation, boldness. Crises have a way of bringing people together, and we have witnessed leaders stepping up and making positive impacts across diverse industries. We hope that they will not soon forget the lessons of leadership forged by the pandemic and that they will all become better leaders going forward as a result.

Furthermore, we also hope that business schools, executive education programs, and corporate training initiatives absorb the lessons leaders have exhibited through this disruption to educate the next generations of leaders who may have not led through it themselves. It is our expectation that the business environment will be fundamentally different in important ways going forward, so those who are training the next generation of business leaders also need to update their curriculum to ensure that new leaders have the skills and mindset to thrive in this changing world. In our previous book on digital disruption, we emphasized the importance of continual learning for both individuals and organizations. That emphasis appears even more critical in a world characterized by disruption.

If we can emerge from the pandemic not only as stronger and nimbler organizations but also as better leaders and people, then we may have truly been shaped by this acute crisis in positive ways. We simply have not been able to do justice to all the stories of inspirational leadership we have observed during our research. As a result, we published a series of profiles in leadership in the *Wall Street Journal* about many of our interview subjects. If you want to know more about the people and stories contained in this book, please search www.deloitte.com/us /transformation-myth or www.profkane.com.

Finally, although we have sought to use this book to encourage leaders and organizations to be optimistic, productively innovate, and inspire through disruption, we also recognize that the COVID-19 pandemic has been the source of hardship for many in our communities. The growth mindset we introduce in chapter 2 is perhaps the single most important aspect of innovating through disruption. Nevertheless, we recognize that no mindset can erase the genuine suffering and tragedies that many have experienced during this acute disruption, whether it be through the loss of loved ones, health, sense of safety, or livelihoods. We want to close by taking a moment to honor and remember friends, family, and colleagues who may have been tragically affected by the pandemic or its broader effects. Having a growth mindset and innovating through disruption does not mean that one also abandons empathy for those who are hurting because of that very disruption.

Acknowledgments

We are big believers in the power of team, collaboration, and connection, and we approached this project with that deep-rooted conviction. While it's impossible to list the many people who have contributed to making this book possible, we want to recognize several groups and individuals.

First, we'd like to thank the dozens of executives who took the time to share their inspiring stories with us. Your insights, authenticity, and leadership brought life to the concepts in this book.

We are grateful to our extended team from Deloitte and Boston College. Many thanks to Rohan Gupta and Joe Greiner for the nights and weekends they spent working through the chapter materials with us. Thank you to Julia MacDonald, Sara Samir, and Emily Walker for the hours you spent pouring through interview materials. Kelly Gaertner, thank you for the many big and small things you did to keep this project on track and for making our article series in the *Wall Street Journal* possible. Thank you to Michael LeFort, Rachel Lebeaux, Abhijith Ravinutala, and Michele Ruskin for your contributions to the *WSJ* interview series and to the many executives who allowed us to feature their stories. We'd also like to thank a number of others who contributed to the development and launch of the book, including: Matt Calcagno, Katie Callow, Lancy Jiang, Dana Kublin, Cat McGuire, Debbra Stolarik, Christine Svitila, Kendall Thurlow, Divya Viswanathan, and Maria Wright.

Thanks to Emily Taber for her editorial guidance and to Paul Michelman for his continued support and belief in this research. We'd also like to thank Deloitte leaders Dan Helfrich, Matt David, Nishita Henry, and Bill Briggs for sponsoring and supporting this effort. And to our colleagues who reviewed drafts and served as sounding boards for our thinking: Ranjit Bawa, Mike Bechtel, Scott Buchholz, Amelia Dunlop, Deborah Golden, Steve Hatfield, Khalid Kark, Anne Kwan, Kristi Lamar, Simon McLain, Nitin Mittal, John Peto, Tom Schoenwaelder, and John Tweardy.

Finally, we'd like to thank the home team: our family and friends who supported us through this process. Thank you for your endless support and encouragement.

Notes

Introduction

1. Clayton M. Christensen, Michael E. Raynor, and Rory McDonald, "What Is Disruptive Innovation?," *Harvard Business Review* 93 (2015): 44–53.

2. Lance Lambert, "Fortune Survey: 62% of CEOs Plan Policy Changes in Response to Current Calls for Racial Justice," *Fortune*, June 18, 2020.

3. See https://www.lexico.com/en/definition/transformation.

4. Gerald C. Kane, Anh Nguyen Phillips, Jonathan R. Copulsky, and Garth R. Andrus, *The Technology Fallacy: How People Are the Real Key to Digital Transformation* (Cambridge, MA: MIT Press, 2019).

5. Khalid Kark, Anh Nguyen Phillips, Bill Briggs, Mark Little, John Tweardy, and Scott Buckholz, "The Kinetic Leader: Boldly Reinventing the Enterprise," Deloitte Insights, May 2020, https://www2.deloitte.com/ch/en/pages/technology/articles/the-kinetic-leader-boldly-reinventing-the-enterprise.html.

6. Andy Cohen and Diane Hoskins, "Insights from Gensler's US Work from Home Survey 2020," Gensler, May 26, 2020, https://www.gensler.com/research-insight/blog/insights-from-genslers-u-s-work-from-home-survey-2020.

1   What the Pandemic Taught Us about Digital Disruption

1. Stanley McChrystal, *Team of Teams: New Rules of Engagement for a Complex World* (New York: Penguin, 2015).

2. Khalid Kark, Anh Nguyen Phillips, Bill Briggs, Mark Little, John Tweardy, and Scott Buckholz, "The Kinetic Leader: Boldly Reinventing the Enterprise," Deloitte Insights, May 2020, https://www2.deloitte.com/ch/en/pages/technology/articles/the-kinetic-leader-boldly-reinventing-the-enterprise.html.

3. Jeffrey Pfeffer and Robert I. Sutton, *The Knowing-Doing Gap: How Smart Companies Turn Knowledge into Action* (Boston: Harvard Business School Press, 2000).

4. Gerald C. Kane, Doug Palmer, Anh Nguyen Phillips, David Kiron, and Natasha Buckley, "Aligning the Organization for Its Digital Future," *MIT Sloan Management Review*, July 26, 2016, https://sloanreview.mit.edu/projects/aligning-for-digital -future/.

5. Scott Berinato, "That Discomfort You're Feeling Is Grief," *Harvard Business Review*, March 23, 2020, https://hbr.org/2020/03/that-discomfort-youre-feeling-is -grief.

6. Bill Gates, "The Next Outbreak? We're Not Ready," TED video, March 2015, https://www.ted.com/talks/bill_gates_the_next_outbreak_we_re_not_ready ?language=en.

7. Alistair Dieppe, "Global Productivity: Trends, Drivers, and Policies," World Bank Group, July 14, 2020, https://www.worldbank.org/en/research/publication /global-productivity.

8. Deloitte Consulting LLP, *Tech Trends 2021*, Macro Forces, December 15, 2020, https://www2.deloitte.com/us/en/insights/focus/tech-trends/2021/macro -technology-trends.html.

9. Val Srinivas et al., "2020 Banking and Capital Markets Outlook," Deloitte, December 3, 2019, https://www2.deloitte.com/be/en/pages/financial-services /articles/2020-banking-and-capital-markets-outlook.html.

10. Andrew Blau, Gopi Billa, and Philipp Willigmann, "The World Remade by COVID-19: Four Planning Scenarios for Resilient Leaders," Deloitte, April 6, 2020, https://www2.deloitte.com/us/en/pages/about-deloitte/articles/covid-19 /four-scenarios-for-business-leaders-world-remade.html.

2   Beyond Surviving: Developing a Digital Resilience Mindset

1. Carol S. Dweck, *Mindset: The New Psychology of Success* (New York: Ballantine Books, 2008).

2. Herminia Ibarra and Aneeta Rattan, "Microsoft: Instilling a Growth Mindset," *London Business School Review* 29 (2018): 50–53.

3. Angela Duckworth, *Grit: The Power of Passion and Perseverance* (New York: Scribner, 2016).

4. Malcolm Gladwell, *Outliers: The Story of Success* (New York: Little, Brown, 2008).

5. Judith Shulevitz, "'Grit,' by Angela Duckworth," *New York Times*, May 4, 2016.

6. David A. Garvin, "Building a Learning Organization," *Harvard Business Review* 71 (1993): 73–91.

7. Gerald C. Kane, "Reimagining Customer Service at KLM Using Facebook and Twitter," *MIT Sloan Management Review*, April 30, 2014, https://sloanreview.mit .edu/article/reimagining-customer-service-at-klm-using-facebook-and-twitter/.

8. "Solutions, Risk Sensing," Deloitte, accessed October 4, 2020, https://www2 .deloitte.com/global/en/pages/risk/solutions/risk-sensing.html.

9. Kapil Tuli, Christopher Dula, and Sheetal Mittal, "Planes, Trains and Social Media," Singapore Management University Case Study, March 30, 2017, https:// store.hbr.org/product/planes-trains-and-social-media/SMU155.

10. Anne Kwan, Maximillian Schroeck, and Jon Kawamura, "Architecting an Operating Model: A Platform for Accelerating Digital Transformation," Deloitte Insights, August 5, 2019, https://www2.deloitte.com/global/en/insights/focus /industry-4-0/reinvent-operating-model-digital-transformation.html.

3   Digital Resilience Readiness: Leading through the Fog of War

1. Derek Thompson, "All the Coronavirus Statistics Are Flawed," *Atlantic*, March 28, 2020.

2. Melissa Repko, "Target CEO Withdraws Forecast: 'None of Us Know How Long This Virus Is Going to Last,'" CNBC, March 25, 2020, https://www.cnbc.com/2020 /03/25/target-scales-back-store-remodels-and-openings-amid-coronavirus-outbreak .html.

3. Warren G. Bennis and Robert J. Thomas, "Crucibles of Leadership," *Harvard Business Review* 80 (2002): 39–45.

4. Claudine Gartenberg and George Serafeim, "181 Top CEOs Have Realized Companies Need a Purpose beyond Profit," *Harvard Business Review*, August 20, 2019, https://hbr.org/2019/08/181-top-ceos-have-realized-companies-need-a -purpose-beyond-profit.

5. "Purpose," Jim Stengel Co., accessed October 4, 2020, https://www.jimstengel .com/purpose/.

6. K. Monahan, T. Murphy, and M. Johnson, "Humanizing Change: Developing More Effective Change Management Strategies," *Deloitte Review*, no. 19, July 14, 2016, https://www2.deloitte.com/tr/en/pages/human-capital/articles/developing-more-effective-change-management-strategies.html.

7. For a more detailed discussion of the importance of trust in driving brand loyalty, see Jonathan R. Copulsky, *Brand Resilience: Managing Risk and Recovery in a High-Speed World* (New York: St. Martin's Press, 2011).

8. Dara Khosrowshahi, "Uber's New Cultural Norms," Uber Newsroom, November 8, 2017, https://www.uber.com/newsroom/ubers-new-cultural-norms/.

9. Chad Storlie, "Manage Uncertainty with Commander's Intent," *Harvard Business Review,* November 3, 2010, https://hbr.org/2010/11/dont-play-golf-in-a-football-g.

10. Novid Parsi, "Communicating with Employees during a Crisis," The SHRM Blog, Society of Human Resources Management, August 28, 2017, https://blog.shrm.org/blog/communicating-with-employees-during-a-crisis.

4   Digital Resilience Readiness: Making Strategic Decisions in the Face of Uncertainty

1. Peter Cohan, "Why Adaptability Helped Some Businesses Survive and Thrive during the Pandemic," *Ink*, August 14, 2020.

2. Ruth Simon, "COVID-19 Shuttered More Than 1 Million Small Businesses. Here Is How Five Survived," *Wall Street Journal*, August 1, 2020.

3. "Amazon's COVID-19 Blog: Updates on How We're Responding to the Crisis," Amazon Day One Blog, July 28, 2020, accessed September 22, 2020, https://www.aboutamazon.com/news/company-news/amazons-covid-19-blog-updates-on-how-were-responding-to-the-crisis.

4. Hugh Courtney, Jane Kirkland, and Patrick Viguerie, "Strategy under Uncertainty," *Harvard Business Review* 75 (1997): 67–79.

5. Daniel Kahneman, *Thinking Fast and Slow* (New York: Farrar, Straus and Giroux, 2011).

6. Thomas H. Davenport, "How to Make Better Decisions about Coronavirus," *MIT Sloan Management Review*, April 8, 2020, https://sloanreview.mit.edu/article/how-to-make-better-decisions-about-coronavirus/.

7. Technical debt (also referred to as technology debt) "is a programming concept that demonstrates the implied cost of ongoing updates or additional

rework by choosing an easy or limited solution that offers quick results over a better approach that would take a longer time to develop"; see https://sitetoolset.com/blog/2020/04/28/what-is-technology-debt/.

5   Developing Your Digital Innovation Superpowers

1.   Traced back to its first appearance in the *Journal of the Society of Estate Clerks of Works* of Winchester, England, in 1908, a version of this parable has appeared in various forms since.

2. Dorothy E. Leidner, Robert C. Beatty, and Jane M. Mackay, "How CIOs Manage IT during Economic Decline: Surviving and Thriving amid Uncertainty," *MIS Quarterly Executive* 2, no. 1 (2003): article 7.

3. Jacob Kastrenakes, "Zoom Saw a Huge Increase in Subscribers—and Revenue— Thanks to the Pandemic," The Verge, June 2, 2020, https://www.theverge.com /2020/6/2/21277006/zoom-q1-2021-earnings-coronavirus-pandemic-work-from -home.

4. Ian King, "Cisco Sees Demand Surge for Webex, Zoom's Larger Rival," *Bloomberg Markets*, March 23, 2020.

5. Nick Statt, "Zuckerberg: 'Move Fast and Break Things' Isn't How Facebook Operates Anymore," C-Net, April 30, 2014, https://www.cnet.com/news/zuckerberg -move-fast-and-break-things-isnt-how-we-operate-anymore/.

6   How to Move at Cloud Speed

1. Nicholas G. Carr, "IT Doesn't Matter," *Harvard Business Review*, May 2003, https://hbr.org/2003/05/it-doesnt-matter.

2. Valerie Lucus-McEwen, "How Cloud Computing Can Benefit Disaster Response," Emergency Management, May 7, 2012, https://www.govtech.com/em /disaster/How-Cloud-Computing-Can-Benefit-Disaster-Response.html.

3. Andrew McAfee, "What Every CEO Needs to Know about the Cloud," *Harvard Business Review* 89 (2011): 124–132.

4. Andrew Winston, "Cloud Computing Is Greener," *Harvard Business Review*, March 2, 2011, https://hbr.org/2011/03/cloud-computing-is-greener.

5. Hemant Taneja and Kevin Maney, "The End of Scale," *MIT Sloan Management Review* 59, no. 3 (2018): 67–72.

6. Nicholas Bloom and Nicola Pierri, "Cloud Computing Is Helping Smaller, Newer Firms Compete," *Harvard Business Review*, August 31, 2018, https://hbr .org/2018/08/research-cloud-computing-is-helping-smaller-newer-firms-compete.

7. Michael Ewens, Ramana Nanda, and Matthew Rhodes-Kropf, "How Cloud Computing Changed Venture Capital," *Harvard Business Review*, October 25, 2018, https://hbr.org/2018/10/research-how-cloud-computing-changed-venture-capital.

7   How to Hyperdifferentiate with Data and AI

1. Alessandro Acquisti, Leslie K. John, and George Loewenstein, "What Is Privacy Worth?," *Journal of Legal Studies* 42 (2013): 249–274.

2. Andrew McAfee, and Erik Brynjolfsson, "Big Data: The Management Revolution," *Harvard Business Review* 90 (2012): 60–68.

3. Beena Ammanath, Susanne Hupfer, and David Jarvis, "Thriving in the Era of Pervasive AI: Deloitte's State of AI in the Enterprise," Deloitte, July 14, 2020, https://www2.deloitte.com/content/dam/Deloitte/cn/Documents/about -deloitte/deloitte-cn-dtt-thriving-in-the-era-of-persuasive-ai-en-200819.pdf.

4. Sam Ransbotham et al., "Expanding AI's Impact with Organizational Learning," MIT-SMR Report on Artificial Intelligence, https://sloanreview.mit.edu /projects/expanding-ais-impact-with-organizational-learning/.

5. Ammanath, Hupfer, and Jarvis, "Thriving in the Era of Pervasive AI."

6. Serdar Yegulalp, "What Is TensorFlow? The Machine Learning Library Explained," InfoWorld, June 18, 2019, https://www.infoworld.com/article/32780 08/what-is-tensorflow-the-machine-learning-library-explained.html.

7. Nitin Mittal and Beena Ammanath, "The Age of With™ Exploring the Future of Artificial Intelligence," Deloitte, accessed October 4, 2020, https://www2.deloitte .com/us/en/pages/deloitte-analytics/articles/exploring-the-future-of-ai.html.

8. Margaret Gould Stewart, "How Giant Websites Design for You (and a Billion Others, Too)," TED video, March 2014, https://www.ted.com/talks/margaret_ gould_stewart_how_giant_websites_design_for_you_and_a_billion_others_too.

9. Jeffrey Dastin, "Amazon Scraps Secret AI Recruiting Tool that Showed Bias against Women," Reuters, October 10, 2018, https://www.reuters.com/article /us-amazon-com-jobs-automation-insight/amazon-scraps-secret-ai-recruiting -tool-that-showed-bias-against-women-idUSKCN1MK08G.

10. Tom Davenport, Jim Guszcza, Tim Smith, and Ben Stiller, "Analytics and AI-Driven Enterprises Thrive in the Age of With," Deloitte Insights, July 25, 2019, https://www2.deloitte.com/us/en/insights/topics/analytics/insight-driven-orga nization.html.

11. Foursquare, "Understanding the Impact of COVID-19 with Foot Traffic Data," March 18, 2020, https://foursquare.com/article/understanding-the-impact -of-covid-19-with-foot-traffic-data.

12. Matthew Hutson, "Artificial-Intelligence Tools Aim to Tame the Coronavirus Literature," *Nature*, June 9, 2020.

8   How to Ensure a Cybersafe Future

1. Blake E. Strom et al., "Finding Cyber Threats with ATT&CK-Based Analytics," MITRE, June 2017, https://www.mitre.org/publications/technical-papers /finding-cyber-threats-with-attck-based-analytics.

2. *The Transformation Myth* is an independent publication and has not been authorized, sponsored, or otherwise approved by Apple Inc.

3. Andrew Burt, "Cybersecurity Is Putting Customer Trust at the Center of Competition," *Harvard Business Review*, March 4, 2019, https://hbr.org/2019/03 /cybersecurity-is-putting-customer-trust-at-the-center-of-competition.

4. Dante Disparte and Chris Furlow, "The Best Cybersecurity Investment You Can Make Is Better Training," *Harvard Business Review*, May 16, 2017, https://hbr .org/2017/05/the-best-cybersecurity-investment-you-can-make-is-better-training.

5. Hal Gregersen, "Digital Transformation Opens New Questions—and New Problems to Solve," *MIT Sloan Management Review* 60, no. 1 (2018): 27–29.

6. Veronica Combs, "Cybercriminals Timed Attacks to Spike during Peak Uncertainty about the Coronavirus," Tech Republic, May 5, 2020, https://www .techrepublic.com/article/cybercrimnals-timed-cyber-attacks-to-spike-during-peak -uncertainty-about-the-coronavirus/.

7. Brenda R. Sharton, "How Organizations Can Ramp Up Their Cybersecurity Efforts Right Now," *Harvard Business Review*, May 1, 2020, https://hbr.org/2020/05 /how-organizations-can-ramp-up-their-cybersecurity-efforts-right-now.

8. World Health Organization, "WHO Reports Fivefold Increase in Cyber Attacks, Urges Vigilance," news release, April 23, 2020, https://www.who.int/news/item /23-04-2020-who-reports-fivefold-increase-in-cyber-attacks-urges-vigilance.

9. Catalin Cimpanu, "Cognizant Expects to Lose between \$50m and \$70m Following Ransomware Attack," ZDNet, May 8, 2020, https://www.zdnet.com/article/cognizant-expects-to-lose-between-50m-and-70m-following-ransomware-attack/.

10. Davey Winder, "COVID-19 Vaccine Test Center Hit by Cyber Attack, Stolen Data Posted Online," *Forbes*, March 23, 2020.

11. Taylor Lorenz, "Zoombombing: When Video Conferences Go Wrong," *New York Times*, March 23, 2020.

12. "Spotting and Preventing COVID-19 Social Engineering Attacks," JP Morgan, May 12, 2020, https://www.jpmorgan.com/commercial-banking/insights/spotting-and-preventing-covid-19-social-engineering-attacks.

13. David Voreacos, Katherine Chiglinsky, and Riley Griffin, "Merck Cyberattack's \$1.3 Billion Question: Was It an Act of War?," *Bloomberg Markets*, December 3, 2019.

14. Zachary Cohen, "State-Backed Hackers behind Wave of Cyberattacks Targeting Coronavirus Response, US and UK Warn," CNN, May 5, 2020, https://www.cnn.com/2020/05/05/politics/us-uk-cyberattack-warning-coronavirus/index.html.

15. Michael Daniel, "Why Is Cybersecurity So Hard?," *Harvard Business Review*, May 22, 2017, https://hbr.org/2017/05/why-is-cybersecurity-so-hard.

16. See https://enterprise.verizon.com/resources/reports/dbir/.

17. James A. Winnefeld Jr. et al., "Cybersecurity's Human Factor: Lessons from the Pentagon," *Harvard Business Review* 93 (2015): 87–95.

18. Michael Parent, Greg Murray, and David R. Beatty, "Act Don't React: A Leader's Guide to Cybersecurity," *Rotman Management Review*, Fall 2019.

19. Greg Bell, "Good Cybersecurity Doesn't Try to Prevent Every Attack," *Harvard Business Review*, October 25, 2016, https://hbr.org/2016/10/good-cybersecurity-doesnt-try-to-prevent-every-attack.

9   Rethinking How We Work

1. David Streitfeld, "The Long, Unhappy History of Working from Home," *New York Times*, June 29, 2020.

2. Marguerite Ward and Shana Lebowitz, "A History of How the 40-Hour Work-week Became the Norm in America," Business Insider, June 12, 2020, https://www.businessinsider.com/history-of-the-40-hour-workweek-2015-10.

3. Vicky Gan, "The Invention of Telecommuting," Bloomberg City Lab, December 1, 2015, https://www.bloomberg.com/news/articles/2015-12-01/what-telecommuting-looked-like-in-1973.

4. Evan DeFilippis, Stephen Michael Impink, Madison Singell, Jeffrey T. Polzer, and Raffaella Sadun, "Collaborating during Coronavirus: The Impact of COVID-19 on the Nature of Work," NBER Working Paper No. 27612, July 2020, https://www.nber.org/papers/w27612.

5. Natalie Singer Velush, Kevin Sherman, and Erik Anderson, "Microsoft Analyzed Data on Its Newly Remote Workforce," *Harvard Business Review*, July 15, 2020, https://hbr.org/2020/07/microsoft-analyzed-data-on-its-newly-remote-workforce.

6. David C. Smith et al., "Workplace Ecosystems of the Future," Cushman & Wakefield, September 2020, https://www.cushmanwakefield.com/en/insights/covid-19/the-future-of-workplace.

7. Elizabeth Dwoskin, "Americans Might Never Come Back to the Office, and Twitter Is Leading the Charge," *Washington Post*, October 1, 2020.

8. "When Everyone Can Work from Home, What's the Office For?," PwC's US Remote Work Survey, June 25, 2020, https://www.pwc.com/us/en/library/covid-19/assets/pwc-return-to-work-survey.pdf.

9. Veronica Melian and Adrian Zebib, "How COVID-19 Contributes to a Long-Term Boost in Remote Working," April 2020, https://www2.deloitte.com/ch/en/pages/human-capital/articles/how-covid-19-contributes-to-a-long-term-boost-in-remote-working.html.

10. Nicholas Bloom, James Liang, John Roberts, and Zhichun Jenny Ying, "Does Working From Home Work? Evidence from a Chinese Experiment," *Quarterly Journal of Economics* 130 (2015): 165–218.

11. Sanna Balsari-Palsule and Brian R. Little, "How to Manage Your Extraverted Employees When You're Working Remotely," *Fast Company*, June 16, 2020.

12. Emily Concannon and Gerald C. Kane, "Social Media Expands Horizons for Workers with Autism Spectrum Disorder," *MIT Sloan Management Review*, June 7, 2015, https://sloanreview.mit.edu/article/social-media-expands-horizons-for-workers-with-autism-spectrum-disorder/?og=Social+Business+Infinite.

13. Christian Jarrett, "The Personalities that Benefit Most from Remote Work," BBC, June 2, 2020, https://www.bbc.com/worklife/article/20200601-the-persona lities-that-benefit-most-from-remote-work.

14. Maryam Alavi and Dorothy E. Leidner, "Knowledge Management and Knowledge Management Systems: Conceptual Foundations and Research Issues," *MIS Quarterly* 25 (2001): 107–136.

15. Mark S. Granovetter, "The Strength of Weak Ties," *American Journal of Sociology* 78 (1973): 1360–1380.

16. Bryan Robinson, "Is Working Remote a Blessing or Burden? Weighing the Pros and Cons," *Forbes*, June 19, 2020.

17. Peter Grant, "Facebook Purchases REI's Unused Hub," *Wall Street Journal*, September 15, 2020.

18. Peter Grant, "Facebook Buys REI's Elaborate New Headquarters as COVID-19 Pandemic Prompts a Sale," *Wall Street Journal*, September 14, 2020.

19. David C. Smith et al., "Workplace Ecosystems of the Future," Cushman & Wakefield, September 2020, https://www.cushmanwakefield.com/en/insights /covid-19/the-future-of-workplace.

20. G. J. Tellis, J. C. Prabhu, and R. K. Chandy, "Radical Innovation across Nations: The Preeminence of Corporate Culture," *Journal of Marketing* 73, no. 1 (2009): 3–23, https://doi.org/10.1509/jmkg.73.1.003.

10   Teaming Your Way through Disruption

1. Elaine Pulakos and Robert B. (Rob) Kaiser, "Don't Let Teamwork Get in the Way of Agility," *Harvard Business Review*, May 12, 2020, https://hbr.org/2020/05 /dont-let-teamwork-get-in-the-way-of-agility.

2. Martha Heller, "Why CarMax's "Shock the System" Digital Strategy Is Working," CIO, April 26, 2017, https://www.cio.com/article/3191884/why-carmax-s -shock-the-system-digital-strategy-is-working.html.

3. Scott D. Anthony and Michael Putz, "How Leaders Delude Themselves about Disruption," *MIT Sloan Management Review* 61, no. 3 (2020): 56–63.

4. Vijay Govindarajan, "Planned Opportunism," *Harvard Business Review* 94 (2016): 54–61.

5. Karl E. Weick, "Educational Organizations as Loosely Coupled Systems," *Administrative Science Quarterly* 21 (1976): 1–19.

11   Rebuilding Disrupted Customer Relationships

1. Charles Duhigg, *The Power of Habit: Why We Do What We Do in Life and Business* (New York: Random House, 2012).

2. Khalid Kark, Anh Nguyen Phillips, Bill Briggs, Mark Little, John Tweardy, and Scott Buckholz, "The Kinetic Leader: Boldly Reinventing the Enterprise," Deloitte Insights, May 2020, https://www2.deloitte.com/ch/en/pages/technology/articles /the-kinetic-leader-boldly-reinventing-the-enterprise.html.

3. James Clear, *Atomic Habits: An Easy and Proven Way to Build Good Habits and Break Bad Ones* (New York: Penguin, 2018).

4. Charles Duhigg, *The Power of Habit: Why We Do What We Do in Life and Business* (New York: Random House, 2012).

5. A. G. Lafley and Roger L. Martin, "Customer Loyalty Is Overrated: Focus on Habit Instead," *Harvard Business Review* 95 (2017): 45–54.

6. Lisa Lacy, "The Pandemic Proved There Is No Brand Loyalty beyond Obsessed Superfans," *Adweek*, August 26, 2020.

7. Jonathan Knowles et al., "Growth Opportunities for Brands during the COVID Crisis," *MIT Sloan Management Review*, May 5, 2020, https://sloanreview .mit.edu/article/growth-opportunities-for-brands-during-the-covid-19-crisis/.

8. "Impact of the COVID-19 Crisis on Short- and Medium-Term Consumer Behavior," Deloitte, accessed October 4, 2020, https://www2.deloitte.com/content/dam /Deloitte/de/Documents/consumer-business/Impact%20of%20the%20COVID -19%20crisis%20on%20consumer%20behavior.pdf.

9. Denise L. Yohn, "The Pandemic Is Rewriting the Rules of Retail," *Harvard Business Review*, July 6, 2020, https://hbr.org/2020/07/the-pandemic-is-rewriting -the-rules-of-retail.

10. Matthew Dixon, Ted McKenna, and Gerardo de la O, "Supporting Customer Service through the Coronavirus Crisis," *Harvard Business Review*, April 8, 2020, https://hbr.org/2020/04/supporting-customer-service-through-the-coronavirus -crisis.

11. Dixon, McKenna, and de la O, "Supporting Customer Service."

12. Robert Shumsky and Laurens Debo, "What Safe Shopping Looks Like during the Pandemic," *Harvard Business Review*, July 24, 2020, https://hbr.org/2020/07 /what-safe-shopping-looks-like-during-the-pandemic.

13. Scott Berinato, "That Discomfort You're Feeling Is Grief," *Harvard Business Review*, March 23, 2020, https://hbr.org/2020/03/that-discomfort-youre-feeling-is-grief.

14. Grant Packard, Sarah G. Moore, and Brent McFerran, "Speaking to Customers in Uncertain Times," *MIT Sloan Management Review*, August 11, 2020, https://sloanreview.mit.edu/article/speaking-to-customers-in-uncertain-times/.

15. Michael Schrage, "Great Digital Companies Build Great Recommendation Engines," *Harvard Business Review*, August 1, 2017, https://hbr.org/2017/08/great-digital-companies-build-great-recommendation-engines.

16. Kian Bakhtiari, "How Will the Pandemic Change Consumer Behavior?," *Forbes*, May 18, 2020.

17. Geoffrey A. Moore, *Crossing the Chasm* (New York: Harper Business, 2014).

18. Michael G. Jacobides and Martin Reeves, "Adapt Your Business to the New Reality: Start by Understanding How Habits Have Changed," *Harvard Business Review*, September–October 2020, https://hbr.org/2020/09/adapt-your-business-to-the-new-reality.

19. L. Keeley, H. Walters, R. Pikkel, and B. Quinn, *Ten Types of Innovation: The Discipline of Building Breakthroughs* (Hoboken, NJ: John Wiley & Sons, 2013).

Index